D1266809

POLITICS AND TECHNOLOGY

THE CONDUCT OF SCIENCE SERIES

Steve Fuller, Ph. D., Editor

Center for the Study of Science in Society
Virginia Polytechnic Institute

The Scientific Attitude, Second Edition
Frederick Grinnell

Politics and Technology
John Street

Politics and Technology

John Street

THE GUILFORD PRESS
New York

© 1992 John Street

Published by The Guilford Press
A Division of Guilford Publications, Inc.
72 Spring Street, New York, N.Y. 10012

Printed in Hong Kong

This book is printed on acid-free paper.

Last digit is print number: 9   8   7   6   5   4   3   2   1

Library of Congress Cataloging-in-Publication Data

Street, John, 1952–
    Politics and technology / John Street.
        p.   cm. — (The Conduct of science series)
    Includes bibliographical references and index.
    ISBN 0–89862–087–2. — ISBN 0–89862–019–8 (pbk.)
    1. Technology and state.   I. Title.   II. Series.
T49.5.S755   1992
338.9'27—dc20   92–3440
                                                  CIP

For my parents, Margaret and Peter Street

Contents

Acknowledgements

This book started life as an undergraduate course, and I owe a considerable debt to all the students who helped formulate the ideas and arguments which follow. My thanks also go to my colleagues at the University of East Anglia: Albert Weale, Tim O'Riordan, Ray Kemp, Alan Cottey and John Greenaway, and friends elsewhere, in particular Mike Stephens, Tim Giles and Ian Forbes. My publisher Steven Kennedy has not only been patient, he has done much to improve my efforts, as has my editor, Keith Povey. Matt Cliffe and Andrew Webster kindly read a draft manuscript, and helped me to clarify my thoughts. Steve Smith has far exceeded the obligations of a good friend and colleague; he has combined criticism with encouragement, and I am very grateful for both. And finally, my thanks to Alex, Jack and Marian Brandon for all sorts of things which reduced the miseries of writing a book and enhanced the pleasures.

JOHN STREET

1 A Few Technicalities

INTRODUCTION

When protesting Chinese students were killed by the army in Tiananmen Square in 1989, their colleagues were able to challenge the official version of events. Despite the strict control over information exercised by the Chinese authorities, the students could reach the world's press and television stations through the use of a FAX machine. When Saddam Hussein invaded Kuwait in 1990, Kuwaiti resistance fighters used FAX machines to give instant accounts of the experience of occupation. Without the technology, such tactics would have been unthinkable. This book is about how such thoughts and actions become possible through the interaction of politics and technology.

There is no escape from either politics or technology. Our daily lives are proof of this. Almost everything we do depends on the technology that surrounds us; and almost everything we say or think incorporates political values and judgements. This book is another example of this state of affairs. Most obviously, it is an attempt to persuade you that we ought to see the world in a particular way, and as such it is an exercise in politics. But it is important to note that my ability to write these words, and yours to read them, depends partly on the vast array of sophisticated technologies used in word-processing, printing, distribution and retailing.

But my purpose here is not just to state the importance of politics and technology; if it were, then the costs in time, paper and technology would hardly make it worthwhile. The point of *Politics and Technology* is not just a reminder of the inescapability of technology. It is to show that its presence, and our dependence on it, raise important political questions. Does the use of technology actually *shape* the kind of political arrangements we have? Can, for instance, democracy only operate in societies which have developed forms of information or communications technology? Is such technology used to thwart or to serve certain political interests? And if technology does play a direct role in politics, who or what determines its form and implementation? In this book, I want to explore the way in which politics and technology become linked: to see how analysis of politics requires reference to technology, and analysis of technology requires reference to politics.

'Technology' encompasses not just nuclear power stations and computers. It extends, for example, to hedgerows, trees and walls. The row of trees outside the American Embassy in London was not planted out of a commitment to natural beauty, but to break up student demonstrations, just as the Paris streets were designed to frustrate revolutionary mobs. The hedgerows of Norfolk, England, also serve the interests of a particular group. They are grown to a specific height so as to help people who enjoy shooting wildlife. A tall hedge forces the birds to fly higher, thereby giving the guns a better target. A simple brick wall can enforce political discrimination. In Cuttesloe in Oxford, until very recently, a large wall blocked an ordinary residential street. It was built with the deliberate purpose of dividing local authority tenants from private home-owners. For the poor tenants, it meant a much longer walk to the shops. It was Oxford's small equivalent of the Berlin Wall. In the home, too, technologies can tell a political story. Take the traditional, pre-electronic piano. In the mid-nineteenth century, the audibility of the piano was increased to allow for its use in concert halls. This modification had a direct impact upon one particular group of players. Michael Chanan records that 'women pianists, who were penalized if they showed too many signs of virility, found themselves disadvantaged, a situation all the crueller because the piano had begun by encouraging not just their self expression but also their professionalisation, in considerable numbers as teachers and accompanists, and a few as virtuosos' (Chanan, 1988, p. 65). What all these examples demonstrate is that even the most mundane technologies have a political dimension.

Some connections between technology and politics are much more blatant, but they too are ignored, not because we cannot see them, but rather because we choose not to look. One of the most telling shots in Claude Lanzmann's documentary, *Shoah*, about the Nazi death camps, is of the mudflap of a truck as it travels down a German autobahn in the early 1980s. The company name on the flap is that of the firm which made the gas for the systematic murders perpetrated by the Nazis forty years earlier. It is too easy to forget that the Nazis' goal – the extermination of the Jews –

required the creation of an almost entirely new technology and bureaucracy to organise it: the bits of paper to authorise transportation, the bureaucratic euphemisms to make genocide seem routine. A political ideology called into being a technology whose sole purpose was to murder people in large numbers. The gas trucks were not just instruments to be used or abused by governments of different political colours, nor were they 'blunt instruments' which one moment might be benign, and the next could be a murder weapon. The gas trucks were a tragic and terrible example of the inseparability of politics and technology.

This book is not, however, an attempt to identify all aspects of the link between politics and technology. Rather, its particular concern is with the relationship between technology and the operation – or reform – of democratic politics. While ideas about democracy continue to abound, there seems to be a gap in the theorising: a reluctance to consider how the theory should be applied to practice. What, for example, does it mean to talk of freedom of speech in a world in which communication is organised through mass communications? The right to stand up in a sparsely attended public meeting hardly seems to be sufficient. But equally, not everyone can have their own radio station or host their own TV show.

My contention is that no student of politics can afford to ignore the way in which technology and political life interact. We need to recognise that technical change can have a dramatic effect upon the way politics is conducted, on the issues that occupy the political agenda, and the interests at play within the political arena. This is as true for information technology as it was for the industrial revolution. It is apparent in arguments about genetic engineering. It can be witnessed in concern over broadcasting standards as cable and satellite TV channels proliferate; or in worries about national sovereignty as new media evolve. The emergence of new issues and interests is not the end of the story. The development of new technology creates new possibilities; it enables people to dream new dreams, or develop new fears. George Orwell's *1984*, Kurt Vonnegut's *Player Piano*, Paul Theroux's *O-Zone*, Ursula LeGuin's *The Dispossessed*, Pete Davies' *The Last Election*, and Marge Piercy's *Woman on the Edge of Time* are just some examples of how imagination and technology can combine. New technology also makes possible new forms of political expression. The burgeoning of the alternative press in the 1960s followed the development of cheaper and easier publishing techniques which allowed changes in style and content. As one editor of the alternative press commented, 'you can't be frivolous in hot metal', but with the new technology you could be (Fountain, 1988, p. 31).

The political scientist's interest in technology extends beyond these examples of the practical impact of innovation. It needs to include the question of how the state deals with its new environment, and perhaps even more crucially how the state actively promotes such developments. Technology does not just sit there; its impact is not simply determined by design. The same technology has a different effect on different societies and cultures, and different technologies emerge from different social settings. We need to think only of how differently television is organised throughout the world.

This, at least, is my argument. There will be those who will disagree with some or all of what I have said so far. *Politics and Technology* is an attempt to justify these claims and to set them in the context of the general concerns of political scientists, and the particular concerns of those who argue about the theory and practice of democracy.

Although it does not constitute a justification of my approach and argument, it might help, by way of explanation for them, if I mention briefly how this book came to be written. At school in the late-1960s I found myself taking science 'A' Levels. My reasons for doing so owed less to a sense of vocation than to a desire to take the course of least resistance – I got better marks for physics and chemistry. My enthusiasm for science derived from an adolescence spent taking apart broken TV sets and making electronic gadgets, and from reading Alexander Fleming's biography. I wanted to invent something. But in the heady, counter-cultural late 1960s, science was decidedly 'uncool', and I began to doubt my commitment to it. These doubts were reinforced by more practical reasoning. Reading Molecular Sciences at Warwick University in 1971, I soon discovered that I was not going to survive as a scientist. I was fortunately allowed to transfer to the Politics Department, where under the patient guidance of my teachers I turned into more of a political scientist than I could ever have hoped to be a molecular scientist. My interest in science, though, never waned completely, and I remained uncomfortable with the stark divide between the two cultures. When I came to teach at the University of East Anglia, I was encouraged to create a course on 'Politics and Technology'. It has been running for a number of years now, and the thoughts and questions it has raised are the basis for this book. My students over this period have made a vital contribution, not only forcing me to make myself and the subject clearer, but also adding their own insights and examples. Together we have, I hope, provided a way of analysing the political causes, character and consequences of technology. This general concern is focused on a specific set of political issues, all of which involve the theory and practice of democracy. Can technology be democratically controlled? How does technology affect the mechanisms of democracy? What conditions does technology introduce into the conduct of democracy? How should technology be assessed in a democratic society?

THE STRUCTURE OF THE BOOK

The structure of the book mirrors this concern. Chapter 2 looks in more detail at the abundant arguments about how technical and political change are to be linked and explained. The essence of my claim is that ideas of autonomous technology or technological determinism, both of which make politics either an appendage to technology or redundant, are distortions of the truth. Instead, it seems more reasonable to conclude that political choices and structures have a definite impact on the form and effect of technology. It follows from this that we need to pay attention to the way in which the state (and other key political institutions) affects the innovation, introduction and impact of technology. This is the task of Chapter 3.

The evolution of modern technology has depended greatly upon the activities of political institutions and political interests, which often set the conditions for the design, development and implementation of technology. The emergence of space exploration and of nuclear weapons are two examples of the way in which political processes have been responsible for technology. What, though, is important is that different political structures and relationships surround different technologies. The variations can be found both within and between countries, and it is on these differences that

Chapter 3 concentrates. Formally defined political processes, however, cannot explain every aspect of technology. They have to be considered in conjunction with science, which provides much of the impetus for technical change.

Chapter 4 examines the degree to which the organisation of, and the interests within, science influence the design and character of technology. If technology is intimately linked with politics, and if science is a crucial factor in determining the nature of the technology, then we need to understand more about what 'science' is and to what extent it incorporates political values and judgements.

At this stage in the argument, it is important not to forget *why* we are concerned with the politics of technology. Chapter 5 acts as a reminder. It examines the different types of effect that technology can have. There are two purposes behind this. Firstly, it simply draws attention to the ways in which technology can affect the lives of people, how it can increase or diminish their life chances, shape their interests or determine their power. In doing so, the chapter identifies the considerations which must be borne in mind when assessing technology. Whose claims should be incorporated into the development and implementation of technology? How can we anticipate the impact of technology, how can we foretell the gains and hazards it will bring?

Chapter 5 forms a vital bridge to the rest of the book, which is concerned with the implications of the arguments presented so far. It is all very well identifying the relationship of technology to politics; the point, though, is to draw out its implications for democracy and the operation of the political system. Chapter 6 sets out the problems which technology seems to pose for democracy. Issues of risk, accountability, technology assessment, together with questions of how technology can be controlled, are linked to different models of democracy to see what difficulties have to be resolved if technology is to be democratically controlled.

Chapters 7 and 8 pursue this theme further by looking at two possible answers to the problems posed by technology. Chapter 7 considers the arguments of the Green movement. The Greens suggest that there is a trade-off to be made between democracy and technology. In order to have popular control over society, it is argued that society cannot be determined by technology. This means that existing forms of both democracy and technology need drastic revision. The Greens' response to the tensions between technology and democracy is not, of course, the only one. Chapter 8 considers the diametrically opposite 'technical fix' approach. This is the argument of those who see technology not as the problem, but as the solution, as holding the key to the realisation of democracy.

Both the Green argument and the case for the technical fix, while resting on key insights, are seriously flawed. Chapter 9, therefore, tries to find a third route by which technology can be democratically controlled. The key to this approach is to avoid the blind faith in, or prejudice against, technology which characterises the other arguments. The answer lies in integrating both democratic theories and practices with forms of technological innovation and application. To give a focus to the argument, it tackles the particular problem of incorporating democracy into mass communications technology.

DEFINITIONS OF TECHNOLOGY

So much for the grand plans and sweeping generalisations. Before we begin the argument in detail, we need to say a few words about technology. (The same might be said about 'democracy', but that is dealt with later.) Many readers, I suspect, will find it hard not to stifle a yawn at the prospect of a (usually inconclusive) discussion of what is meant by a particular term or definition. Such exercises have little more literary appeal than accounts (such as the one above) of what the author plans to do. It is not that definitions are unimportant; any discussion depends on some shared understanding of the terms used by the discussants, but I suspect that such an understanding emerges more clearly through illustration and the general debate, than through any deliberate attempt to fix the terms in advance. Nonetheless, it is necessary to set the broad outlines of 'technology', to identify some of the key ideas that people have in mind when they use the word.

Technology is typically thought of as a piece of hardware employed or fashioned to serve a particular purpose. It is a nuclear power station or a car. The initial assumption, then, is that technology is constituted by human-made objects which have instrumental value. While it is true that we talk of the human body or of the natural world as technically complex, we do not mean that lungs or brains are pieces of technology. Rather we are using technology as a metaphor. The brain, we choose to think, behaves like a computer, or the heart like a water pump, and so on. But they are not the product of human design and construction, although computers and heart pacemakers clearly are technologies. At the same time, a technology does not need to be 'a machine' in the conventional sense. The pencil is a technology; so too is a chemical used to fertilise the land or prolong the life of food. The technology does not itself have to be artificial (that is, manufactured); it may be naturally occurring. What determines its status as 'technology' is the deliberate and conscious use of it by human agents. This general definition conflates the distinction Marx made between a tool and a machine. For Marx, a tool was something controlled by its human user, without whom the tool could not be operated. A machine, by contrast, did not depend on the human user, either for its power or its operation (Marx, 1954, p. 351–5). There is obviously something in the contrast, but for our purposes there is no need to make this the basis of our definition of technology. The difference between a tool and a machine serves better as a way of thinking about technology: what sort of control does the user of, say, a word-processor have? This is a more important question than: is the word-processor a tool or a machine?

At the same time, 'technology' does not just refer to the physical form, the pieces of metal, the electronic components, the chemical compound. Technology refers to the way in which the parts are organised, through the application of knowledge, to realise their particular purpose. This broadens our understanding of technology quite considerably. Firstly, it draws in the principles which enable the components to have their effect. Computer technology, for example, is the science of semi-conductors, the ability of some materials to allow electrons to flow in one direction, and their potential as switching devices. Nuclear power stations are the embodiment of a wealth of knowledge about radioactive decay, electric current generation, and much else besides. As the advertisement for Zanussi washing machines used to say, technology is 'The Appliance of Science'. More accurately, perhaps, 'science' is an umbrella term for all aspects of human knowledge. ('The appliance of all aspects of human knowledge' does not, I grant, have the sort of ring that would appeal to an advertising copywriter.)

The second way in which our notion of technology is broadened is by including the structures which enable it to operate within society. A complex piece of mechanical engineering only becomes 'a car' when there are roads and regulations which enable it to operate as a means of transport; otherwise it is a museum piece. People have to be organised, as well as scientific principles applied, for a technology to have a proper existence.

A fire in a London underground station, in which thirty-one people were killed, brought home the complexity of even the most familiar of technologies. The system by which millions of people were delivered to their destinations depended on the coordination of countless different mechanical and human elements. The fire demonstrated the fragility of the various parts of the structure. As James Dalrymple reported,

Fire safety had become sloppy throughout the system. There were fire extinguishers hidden behind temporary panels. There were water buckets that had not been filled for years. There were hosepipes hidden in long-forgotten places, missing vital parts. Some staff did not even know where to find fire-fighting equipment. There were ancient, crotchety water-fog systems that had never been tested within memory, with valves locked in their own dust. The controls for this apparatus were inside the chamber where the fire they were meant to control would be raging (*The Independent*, 25 June 1988).

The fire showed that the technology – the underground railway – was made up of a set of structures and relationships, each of which incorporated a set of judgements. These represented views about, for example, the relative importance of cost, speed and safety.

'Technology', therefore, is not just the hardware, nor is it just the set of arrangements which enable that technology to operate; it is also a set of decisions about how that technology ought to work. Arnold Pacey summarises this extended view of technology when he talks of 'the application of scientific and other knowledge to practical tasks by ordered systems that involve people and organizations, living things and machines' (Pacey, 1983, p. 6). Daniel Bell also embraces this broad definition of technology: 'the organisation of a hospital or an international trade system is a *social* technology, as the automobile or a numerically controlled tool is a *machine* technology. An *intellectual* technology is the substitution of algorithms (problem-solving rules) for intuitive judgements' (Bell, 1973, p. 29; his emphasis).

POLITICS AND TECHNOLOGY

In identifying these very basic features of technology, we can see the connection between it and the typical concerns of social scientists. Technology seems to both shape and reflect the type of society we live in. Such a thought underpins the interpretations put upon Marx's claim that 'Technology discloses man's mode of dealing with Nature, the process of production by which he sustains his life, and thereby also lays bare the mode of formation of his social relations, and of the mental conceptions that flow from them' (quoted in Rosenberg, 1981, p. 9). Our technology enables us to realise certain wishes, and in doing this we change our conception of ourselves and the world. Think of the development of transport technology. As the cliché has it, travel broadens the mind. It does much more besides. Roads have to be built, regulations introduced, industries created, and so on. The decisions which contribute to this infrastructure set the agenda for 'transport'. And in the process, a bias is created in favour of certain questions and issues; in particular, how to improve, or cope with, the system that is already in place. At the same time, the general political agenda is required to accommodate new political issues: road deaths, pollution, motorway routes, and so on. Not all technologies have so extensive an effect, but those that do are sometimes singled out as 'base technologies' (Jamison, 1989) or 'defining technologies' (Bolter, 1986). They contribute, it is said, to the basic political character of society.

Because technology is not simply a matter of hardware, the emergence of, or change in, technology can be expressed through new relationships. The liberation provided by a technology is not cost-free. It may, for example, be accompanied by an additional element of dependence. If your word processor breaks down, you have to send the machine back to the manufacturer; if your pencil breaks, you sharpen it. Electric aerials on cars are easy to raise and lower, but if they go wrong, you cannot listen to the radio.

While technology may reduce freedom as well as increase it, we need to recognise that one implication of this general account of technology is that it is constituted by a set of choices. Because any technology combines a number of elements – hardware, bureaucracy, and so forth – they can each be accorded different priorities. These are based on political judgements and may draw on a wide range of criteria. In the summer of 1988, British newspapers were filled with stories about the air passengers who were stranded at airports, waiting for flights to continental Europe. Much of the blame was focused on the air traffic controllers, and the 'solution' was seen to lie in relaxing the regulations which prevented night flights. Arguably, though, the real source of the problem lay with the fact that, while airlines and holiday firms were encouraged to create more demand for package holidays, very little attention was given to the infrastructure needed to service this new market. The character of air travel was shaped by decisions about which aspects of it were 'important'. Michael Goldhaber writes: 'Technology is a human activity. Far from being apolitical, the technology that gets developed is a direct result of political choices, choices that could be made differently' (Goldhaber, 1986, p. 4). This is evident in the safety elements which are (or are not) built into the planes carrying tourists to their resorts. The seats face forwards and there are no smoke hoods provided, despite the strongly argued claim that both would improve passenger safety. The airlines, though, worry about their passengers' reluctance to travel backwards and about the cost of the hoods.

This suggests a second, and perhaps overlooked, feature of technology. It cannot be understood simply in terms of its functions. There are the passions it arouses, satisfies and frustrates. People care about technology, about how it looks, about what it *means* to them. It is nearly impossible to be indifferent about technology. We invest it with all kinds of meaning and symbolism, which is then often used by advertisers and film-makers. Technology can be made to evoke a whole range of emotions – it can represent control, or efficiency, or perfection, or cold inhumanity. Car advertising, for instance, works with many of the ideas we have about technology: '*Vorsprung Durch Technik*', as the Audi people say. It would, of course, be wrong to confuse an advertiser's strategy with a consumer's reality. But while the consumer might understand a technology quite differently from the copywriter, the point is that a meaning is being attributed to it. We cannot separate culture from technology. As Neal Ascherson reflected in the opening days of the 1991 Gulf War, military technology is not simply a means to an end; it also embodies a sense of 'strength':

> The Maxim gun did not guarantee press-button victory in the 1890s, any more than the bomber biplane in Arabia between the wars or Stealth, Tornado and Cruise today. And yet, when they seem momentarily to fail, the First World is convulsed by doubt which is not just military but cultural, as if collective virility had failed (*Independent on Sunday*, 20 January 1991).

The way things look or seem may be as important as what they do. The Greens have strong views about cars. They want them eliminated or heavily constrained. Other people have strong views about cars too. They, for example, mistrust drivers of certain cars; they would not be seen dead in a Volvo; they think a Fiat Uno is 'them'. Peter York (1985), in dissecting modern life-styles, argues that technology helps to shape and define those styles. In discussing the emergence of space technology, Peter Marsh makes a similar point: 'It would be wrong . . . to think of developments in space technology as the result simply of initiatives by self-seeking governments or by greedy commercial organizations. There is also a spiritual side to space exploration' (Marsh, 1985, p. 10).

Political scientists do not usually pay much heed to such features of life; nor do students of technology. Or rather they do not in their professional life. As consumers and citizens, I suspect, they do. Whatever is the case, there seems to be an argument for giving more weight to technology's cultural dimension. Understanding technology and its political importance may involve more than comprehending the interests or the purposes that it serves. These are clearly important; technology also plays a key role in the organisation of the political economy of the country. But this in turn depends upon the way it is integrated into the understanding individuals and groups have of themselves and the world they inhabit.

Albert Hirschman shows how the experience of technology is linked to culture and political action. His argument begins with the recognition that consumer durables – refrigerators, cars and so forth – provide comfort rather than pleasure. Or rather the pleasure is only derived once, in the act of first acquiring the good; after that it is taken for granted (Hirschman, 1982, p. 32). The result is that consumers experience a sense of disappointment; they expect a regular pleasure that is not forthcoming. Instead, they move from the state of discomfort – food that goes bad, the need to walk or depend on public transport – to the state of comfort provided by the fridge or car. The frustration with the limits to pleasure, argues Hirschman, lies behind the change in car design and technology. 'Many consumers', he writes, 'strive to fight the disappointment that they sense will be connected with the purely utilitarian aspect of the car by spending more on the vehicle than is strictly necessary for utilitarian purposes' (Hirschman, 1982, p. 36). It is this that explains their tolerance of 'built-in obsolescence'. But for Hirschman the experience of disappointment also has the more generalised, if not permanent, result of pushing people away from private consumption towards public action (ibid., Ch. 4). One expression of this is the rise of the Green movement.

The importance of Hirschman's argument lies in his recognition of the political significance to be attached to use of, and reaction to, technology. Technology is invested with meaning and expectations, and any account of its role in modern society must recognise the implications of this process. The effect of technology on the way we live is partly determined by the images, ideas and practices which are incorporated in it. It is not just that political interests and technology are linked, nor just that political and technical change affect each other, but that contemplating politics and contemplating technology are part of the same process. This book is an attempt to show how.

2 Political Change and Technical Change

INTRODUCTION

Imagine some future world. How would you begin to describe it? For many people, the easiest solution is to focus on the technology available in such a place: the labour-saving devices, the means of oppression, the cultural choices, the communication networks. Not everyone would think like this, but most would. It is, after all, the way our popular culture often works. In futuristic adverts, in science fiction films or novels, the world is evoked through the technology. Such shorthand taps an important intuition: social and political change is marked and moved by technical change. If this intuition is accurate, then the implications for the study of politics are obvious. No student of politics can afford not to be a student of technology.

But does technology actually generate political change? Certainly it is not uncommon to find politicians who claim that the development of technology is crucial to securing economic growth and social well-being. It is not hard to see why this view carries some weight. At a superficial level, it accords with the way we often talk about the world, the way certain kinds of change seem 'inevitable', the way we embrace 'progress'. It also fits with the way we attribute 'responsibility' to technology. Following the gales that devastated the woodlands of Britain in 1987, the blame for failing to predict the storms fell not on the meteorologists but on their computers. It fits too with the idea that if something is technically possible, then its implementation is inevitable. President Reagan's adoption of the Strategic Defense Initiative (SDI), for example, was described as the playing out of a technical 'logic'. Because it was possible, the argument ran, then it had to be tried. This story can be made to fit a general account of US military strategy. The technology determines the political rationale. Strategy is devised to cope with the existence of new weapons, rather than the former giving rise to the latter. Carl Friedrich von Weisacker recalls how in Germany in 1939 he felt

14

the irresistible pull towards the construction of the atomic bomb. As he wrote later: 'At the time [1939] we were faced with a very simple logic . . . [T]he Atomic bomb exists. It exists in the mind of some men. According to the historically known logic of armaments and power systems, it will soon make its physical appearance' (quoted in Rhodes, 1988, p. 312). The change in technology was 'inevitable', and the politics simply followed on its coat-tails. To understand political change, the implication seems to be that we need to understand technical change.

But it is possible, reading the same quotation from von Weisacker, to draw a quite different conclusion. Latching onto the phrase 'power systems', it could be argued that these were what really determined the development of the bomb. The counter-claim is that it is not politicians who are the slaves of the technologists; it is the other way round. SDI was begun not because of the logic of the technology, but because of President Reagan's ambitions and because of the political interests organised around the development of military technology.

There are, in short, at least two quite opposite interpretations of the relationship between politics and technology, one which says that technology determines politics, and the other which says politics shapes technology. The dilemma posed by these alternatives is nicely captured by Lionel Trilling, who asks whether ideas of 'self' generate the technology of mirrors, or whether it is the other way round:

> The French psychoanalyst Jacques Lacan believes that the development of 'Je' was advanced by the manufacture of mirrors: again it cannot be decided whether man's belief that he is a 'Je' is the result of the Venetian craftsmen's having learned how to make plate-glass or whether the demand for looking-glass stimulated this technological success (Trilling, 1972, p. 25).

These competing claims represent the extremes of a complex and abstract debate. To go some way towards mitigating the abstraction and illustrating the complexity, consider a further example: the treatment of the man often credited with pioneering the development of the computer, Alan Turing.

In 1952, Alan Turing was charged with 'gross indecency'. His offence had been to have sex with another man. Before the law changed in the 1960s, in Britain all homosexual contact was illegal. There was no allowance for consenting adults over twenty-one in private. Turing pleaded guilty and was put on probabation. A particular condition, however, was attached to the judgment. Turing had to 'submit for treatment by a duly qualified medical practitioner at Manchester Royal Infirmary' (Hodges, 1985, p. 472). He was given, as a result, organo-therapy, which meant a course of injections with hormones which had unpleasant side-effects. Homosexuality was treated as a medical condition, as an 'illness' caused by an hormonal imbalance.

Did Turing's treatment illustrate the way in which a technology – the hormonal treatment – shaped the judge's decision? Or did medical and judicial interests combine to produce a definition of homosexuality that made a technical solution necessary? This chapter assesses these competing claims. It does so by examining, firstly, the argument that technology in some way determines political processes; it then considers the opposite argument, the idea that the key determinants of technology are political choices or interests. Finally, the chapter ends by attempting a synthesis of these two extremes, in which politics and technology relate dialectically, each influencing the other. Before any of this, however, we need to look more closely at a central feature of the relationship between technology and politics – technical change itself. If technical change shapes political processes, or if politics determines the form of technical change (or if there is some variant of the two), we need to know what is meant by 'technical change'.

DEFINING TECHNICAL CHANGE

A simple model of technical change focuses on two stages: (1) innovation and invention, and (2) implementation and regulation. Change is about both the creation of technology and its introduction into society.

Change as innovation and invention

The most familiar form of technical change is innovation or invention. This may involve the creation, by whatever means, of a new way of performing an old task or process; or it may mean the development of an entirely new product or possibility. Traditionally, such breakthroughs are attributed to an inventor: James Watt and the steam engine, Alexander Bell and the telephone. The validity of such attribution is often challenged. It is observed, for example, that there are instances of simultaneous invention or of disputed claims to the original breakthrough (the invention of television is one example, see Wheen, 1985, pp. 11–39). There are instances, too, of the same technology emerging at roughly the same time in quite different places (the computer emerged in the US and UK almost simultaneously). Such moments, though, do not affect the general contention that technical change can take the form of creative leaps by individuals or groups of individuals.

The idea of invention is, however, challenged by those who doubt the very notion of innovation and the attribution of individual responsibility for it. At one level, the argument here is that technical change is actually a series of incremental developments, each dependent on its predecessor. To describe these increments as innovations is to overlook the *process* which links them. Their similarities are, in other words, more important than their differences. The consequence of such an argument is to play down the role of the individual inventor or innovator. They become mere cyphers of the general process. Although the scientists Crick and Watson unravelled the double helical structure of DNA, and were rewarded with the Nobel Prize for their efforts, there are several reasons for qualifying their importance. Firstly, their breakthrough was based on work being done by others, in particular the X-ray work done by Wilkins and Franklin at King's College, London. Secondly, it seems almost certain that had Crick and Watson not uncovered the structure of DNA, someone else would have – and within months of their having done so. The most likely candidate was the US scientist Linus Pauling, who was prevented from seeing the X-ray photographs of DNA because he was denied a visa by the British immigration officials. Had he been able to see the photographs, he might well have beaten Crick and Watson. This speculation is premised on the idea that innovation owes less to individuals than to ways of thinking, more to processes than to discrete inventions.

But should we be won over by such objections to the 'invention' model? We might not wish to make too much of the role of specific individuals, but such a move should not exclude the role of other factors which no systematic account could incorporate. It would be a mistake to reduce all technical change to a pre-determined process. The detailed examination of technical change reveals that it often entails observations, leaps of imagination or mere serendipity which defy prediction (Hughes, 1983, Ch. 3). The particular timing and character of each step in the chain of technical development makes it unique, requiring a specific explanation. Furthermore, there are two dimensions to that explanation: the internal processes of innovation and the external pressures for change.

The internal account focuses on the thoughts and actions of those responsible for the particular forms and character of the technology. It tells the story of the scientists and engineers who have the knowledge and skills to make the technology work. It also tells of the relations between the professionals, their disciplinary codes and their institutional interests. The external account dwells upon the wider context of the development of technology, the social, economic and political interests which generate the incentives or needs for the technology (Jamison, 1989, p. 512). The two perspectives are not necessarily incompatible. Rather they focus on different aspects of technical change, and ask different questions about how it happened. An internal account of the evolution of the telephone focuses on the technical and scientific processes that made possible the conversion of sounds into electric pulses; an external account concentrates on the drive for new forms of communication.

One point needs to be emphasised in this discussion of technical innovation and invention. Not all technical change takes the form of a 'breakthrough'. Many are modifications of existing technologies. There is, of course, a real difficulty in determining what constitutes a modification and what a breakthrough. Is the compact disc (CD) player, for example, just a more sophisticated version of the traditional record turntable? Is the microchip just a smaller, cheaper version of the printed circuit? The answer is, I suppose, 'yes' and 'no'. The CD player is obviously part of a family of technologies for reproducing recorded music. At the same time, the method by which it does this is novel: the replacement of the stylus by the laser beam. It is this that represents technical change, but it is confined within an established technology or function. On the other hand, what the CD also represents is the ability to reproduce sound digitally. This has enormous significance for the capacity to make and manipulate sound; not only can perfect copies be made, but the original sounds can also be transmitted over large distances with no loss of quality. CD represents the beginning of a new form of technology, one that breaks with previous traditions for sound creation and reproduction. So in one technology we can find both modest modification and major innovation.

Change as regulation and implementation

Technical change is not just marked by invention and innovation. It is also distinguished by implementation, the decision or process by which a particular innovation is introduced into society. The gap between innovation and implementation varies with the type of technology and the environment. There is no simple generalisation to be made. Wind-powered generation is a technical innovation which in many countries remains at the prototype or experimental stage, whilst in others it is an active and viable source of power.

The implementation of technical change also involves its regulation. A technology can be changed, not by anything intrinsic to its design or internal workings, but by the uses to which it is put or by the regulations applied to its operation. Decisions to deregulate broadcasting, for example, change the uses to which communications technology can be put. Equally, uses can be discovered by users, and these may owe nothing to the manufacturer's intentions. Such moments constitute a case of technical change. This has occurred with many innovations in the world of popular music. The development of the cassette recorder, for instance, helped to change the conditions of work for musicians, the business practices of the record industry, and consumption of music by fans. Many of these uses were pioneered by the users not the makers of the technology.

In defining technical change, therefore, we need to be aware of the different levels at which change happens. Not only do we need to distinguish between the internal and external accounts, we also need to focus on the different forces pushing for, or generating, change. Words like 'innovation', 'invention' and 'breakthrough' have to be used cautiously.

THE DIRECTION OF TECHNICAL CHANGE

Much of the previous discussion of technical change assumes that it is measured by the introduction of new devices or processes. But this is not all that is involved. We need to ask what 'new' means; we have to examine what the novelty consists of, where it takes the technology. Technical innovation does not occur randomly; it follows identifiable paths and trends. There are two familiar ways of characterising the direction of technical change. Each typically entails a clear judgement. It is either 'progress', a change for the better; or 'regress', a change for the worse.

Technical change as progress

According to this view, each innovation in a technology constitutes an improvement on its previous form. As Bell writes, 'technological progress consists of all the better methods and organisations that improve the efficiencies (i.e. the utilization) of both old capital and new' (Bell, 1973, p. 191). Technical change, therefore, takes the route marked by systematic improvement.

There are different explanations advanced as to why technology should progress in this way. One view is that it represents the realisation of science, revealing itself in ever increasing control over nature. Human vulnerability to nature is replaced by mastery of it, and with this comes the realisation of freedom (as the absence of dependence). Alternatively, the pattern of change may be attributed to competition, by which the incentive to change is fuelled by the desire to survive in the market. This view can be discovered in Schumpeter's notion of creative destruction in *Capitalism, Socialism and Democracy* (Schumpeter, 1976, Ch. 7). Entrepreneurs push technology by exploiting each innovation, thereby moving technology closer to 'perfection'. Whether technical change is driven by competition or human development, it is claimed that technical change serves to improve the quality of life.

Technical change as regress

Technical change can be regressive in one of two senses. Firstly it can be seen to diminish the quality of life, to limit human freedom or happiness. This view is commonly associated, if not always correctly, with the Green movement. With such a perspective, technical change will be resisted because it threatens established ways of life. The Amish community in the US has, for instance, chosen to live without many of the technical innovations of the twentieth century. Iran, too, in the aftermath of the Shah has rejected many elements of modern Western technology. In many critiques of society, from the left and the right, technology is portrayed as oppressive, as threatening to standardise or homogenise society, diminishing freedom and individuality.

Such views see technology as politically regressive. But technology can be seen as regressive in other, less pejorative ways. Take the example of the Walkman personal stereo. Shuhei Hosokawa observes:

> It is interesting that, from the technical and technological point of view, the 'progress' from the portable radio-cassette or car radio to the Walkman is very minor, contrary to the conspicuous transformation on the level of praxis. . . . It is technologically a simpler object [than the cassette recorder]. . . . It represents functional reduction, technological regression (Hosokawa, 1984, p. 168).

Viewed like this, the Walkman is technically, not politically, regressive. But there are those who would argue that the Walkman has further privatised the world, cutting people off from each other. Such a reading makes the Walkman politically regressive.

The point here is not to offer a definitive account of the direction of technical change. Rather it is to show how conceptions of change may be value-laden, and that the same change may be viewed quite differently. There is, in short, no simple or uncontentious way of labelling change. This is not exclusive to personal stereos. Many technologies combine these ambiguities, enabling innovation to mean both a novel development and a refinement of past achievements. The conclusion I draw from this is that we do not want to attach great importance to ideas of 'progress' or 'regress'. Their interest stems more from the fact that they are applied ideologically, rather than from their descriptive value. If someone talks of 'progress', it is worth asking what they mean, what they see as having improved. Otherwise, for analysis of technical change, it is more important to ask how change is organised than to establish whether it counts as progress.

UNDERSTANDING TECHNICAL CHANGE

In his *Networks of Power*, a history of the electrification of Western society, Thomas Hughes (1983) offers a persuasive picture of technical change. His model begins by echoing the simple two-stage development that we described earlier; 'innovation and development' is followed by 'technology transfer'. But Hughes not only adds a number of further stages, he also enriches the portrait of these opening ones.

The key idea for Hughes is the systemic character of technical change, during which operate a myriad of interconnecting relations. The development of a major technology draws in an ever-increasing number of interests – from the inventors and entrepreneurs at the beginning, to the financiers and politicians involved in the implementation. These professional interests are tied to a technology whose operation constantly raises new applications and problems, and thereby draws in more interests and demands. There is constant pressure to meet these challenges, to make the technology work, and in the process the technology itself grows and changes. It takes on the form of an evolving system with a momentum of its own. David Landes captures the spirit of this idea when he writes that in the movement towards technological improvement, 'change begat change':

> For one thing, many technological improvements were feasible only after advances in associated fields. The steam engine is a classic example of this technological interrelatedness: it was impossible to produce an effective condensing engine until better methods of metal working could turn out accurate cylinders. For another, the gains in productivity and output of a given innovation inevitably exerted pressure on related industrial operations. The demand for coal pushed mines deeper until water seepage became a serious hazard; the answer was the creation of a more efficient pump, the atmospheric steam engine (Landes, 1969, pp. 2–3).

Rather than talking of progress or regress, such accounts are best understood as a case of what Hughes calls 'reverse salience'. Technical change is essentially an exercise in problem solving

(Hughes, 1983, Ch. 4). The development of the technological system depends on correction of dysfunctions and bottlenecks that occur with its application and expansion. Technologies do not come about as the expression of some grand design, but as the result of constant modification and adaptation.

Viewed in this way, technical change is automatically linked to political change. The ability to meet the problems posed by technologies does not just require technical skill; there are organisational and administrative problems to confront. The question is, though, how is this link to be understood? Does the technology force change upon political institutions, or do those institutions demand technical change? The next section is an attempt to answer this question.

POLITICAL CHANGE AND TECHNICAL CHANGE

We begin by considering two theories which claim that technology determines political change. The first talks of 'autonomous technology'; the second of 'technological determinism'. Although the two contain some common features, it makes more sense to consider them separately. We then reverse the causal chain, and consider the argument that politics determines the form of the technology.

Autonomous technology

The theory of autonomous technology claims that technology acquires an independent momentum, which not only puts it beyond human control but which also allows it to order all human activity, including politics. The account I give below does not belong to any one writer, but is synthesised from, among others, the writings of Jaques Ellul (1964) and Herbert Marcuse (1941 and 1968).

If technology is autonomous, choice and judgement play little part in the direction in which society is moved; one technical advance calls into existence another, and so on. It is a process that seems to have its own logic and its own driving force. When he was Britain's Minister for Information Technology, Kenneth Baker was fond of saying that the choice between adopting information technology and ignoring it was the same as the choice between automation and liquidation. It was no choice at all. Technology creates a relentless and constant pressure for change, but affords no opportunity to decide how. The experience of competition can often take this form. For individuals, competition can be a constant drive to improve their performance without their being able to determine the direction and pace of that change – each rival simply works under the injunction: 'do better'. Looking back, they can see that their world has altered dramatically, but they cannot claim that they have controlled their destiny. They have been caught up in the competition. They have been changed by it, but they have had little choice, and have exercised no control. They have been running just to keep up.

Autonomous technical change occurs independently of outside influence. Technology appears to have a mind and momentum of its own, and as such is thought to be unstoppable. The direction of change is set by a logic, or rationale, which is peculiar to technology.

In themselves these points about technology might have little significance were it not for a further dimension to the argument. Autonomous technology is not just a thesis about technical change but also about social change. What motivates the theory of autonomous technology is the thought that the peculiar logic of technology is extended to all aspects of social life; it dictates the terms on which that life is lived. The dominance of technological logic is manifested in the elimination of difference, as everything, including the state, is brought under its influence. As Ellul writes:

> the structures of the modern state and its organs of government are subordinate to the techniques dependent on the state. If we were to consider in turn each of the indispensable services of the modern state, we would find that they are becoming more and more alike, regardless of the theory of government under which they operate (Ellul, 1964, p. 271).

For Ellul, there is no way of stopping this path towards the similarity of states. The state can only serve the functions that technological rationality allows. The idea of autonomous technology can be represented as follows:

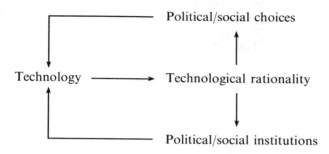

This process does not end with the state. The private world of personal intimacy has been colonised by technological rationality. Sex and personal relations are a matter of technique rather than feeling. A magazine launched in Britain in 1988 is typical. *One to One*, in several monthly instalments and with many helpful illustrations, explains how to run your love-life. The tone and the style of *One to One* is hardly different from those of the DIY, cookery and motor maintenance magazines stacked next to it. It treats relationships as things to be fixed like a flat tyre or assembled like a fitted cupboard.

Those who write of autonomous technology are not simply observing its impact, but are criticising it too. Technological rationality is a limited and impoverished thing which excludes crucial elements of human experience. The obsessive concern with technique is seen to undervalue expressive reasoning to the benefit of instrumental – or technological – reasoning. The autonomous technologists are concerned that the only values that operate in society, and by which actions or policies are judged to be good, are those of technical efficiency: economy, speed, and so forth. This, in turn, legitimates the extension of technology into yet new areas of life. And so the technological screw twists tighter. As Marcuse observed (1968, p. 19): 'A comfortable, smooth, reasonable, democratic unfreedom prevails in advanced industrial civilisation, a token of technical progress'. Such a view is not exclusive to the 1960s. It was still being expressed in the 1980s; this time it was set against the background of the growth of computer technology. Joseph Weizenbaum, a disillusioned computer scientist, reported that the reliance on computer systems has had 'two important consequences':

First, decisions are made with the aid of, and sometimes entirely by, computers whose programmes no one any longer knows explicitly or understands. Hence no one can know the criteria or rules on which such decisions are based. Second, the systems of rules and criteria that are embodied in such computer systems become immune to change, because, in the absence of a detailed understanding of the inner workings of a computer system, any substantial modification of it is very likely to render the whole system inoperative and possibly unrestorable. Such computer systems can therefore only grow, and their growth and the increasing reliance placed on them is then accompanied by an increasing legitimation of their 'knowledge base' (Weizenbaum, 1984, pp. 236–7).

Human agency is automated by the impact of technology and its rationale.

There are a number of elements in the autonomous technology thesis, and it will pay to separate them if we are to analyse the argument effectively. A key element is the notion that technology has a rationality of its own. It is this idea which sustains the notion that technology can act independently of human agency. Were there no internal logic to technological change then any explanation of that change would have to draw on other factors; technology would no longer be autonomous and the thrust of the argument would be lost.

The idea that technology has an internal logic can be imputed from the claim that whoever works on the development of technology, the result will be the same. Such a theory underpins the observation of simultaneous inventions. This evidence is used to suggest that technology develops independently of the people engaged in its propagation.

The second stage in the argument focuses on the way technology interacts with society. We might concede that technology does have an internal logic, but it does not follow from this that it moulds social life in its own likeness. The claim that it does rests on two ideas. The first has to do with the way technical solutions establish superiority over other types of solution. When faced with a problem or decision, we want the best solution. There is nothing very controversial in this, of course, until we ask what counts as the 'best'. The 'best' must not be measured subjectively; it cannot be merely a matter of whim. There has to be some standard of comparison, some way of evaluating two or more possible solutions. This is vital for public, political decisions. When we are dealing with a large, complex society, and when the solutions require the commitment of large public funds, a government has to be able to identify its chosen course of action as the best available. Otherwise its decisions will appear arbitrary or irrational. In Margaret Thatcher's famous words, government will want to claim that 'there is no alternative' to its policy. What governments (or any authority) are looking for is a grounding for such claims. Technology seems to offer one by setting objective standards of comparison which allow for unbiased judgement about the relative merits of competing schemes.

The key idea is efficiency. If the task of a transport system is to get people from point A to point B, then the scheme that does this fastest is best. If the task of an electric grid is to provide electricity, then the power stations that do this most cheaply, for any given set of customers, is the best. If someone offers a form of electricity generation that is 'too cheap to meter', as was claimed for nuclear power, then it is hard to resist the development of that technology. Certainly, if one's reason for objecting is that the 'quality' of life will be adversely effected, or if freedom will be eroded, then such arguments are difficult to sustain in the face of the facts. No two people may agree as to what counts as an erosion of liberty or what contributes to the quality of life, but everyone can agree that two is larger than one. In his novel *The Tin Men*, Michael Frayn pursues this argument to its 'logical' conclusion. Football is just a device for getting random results which can then produce winners for the Football Pools. It is time-wasting and costly having twenty-two people chasing a football on a Saturday afternoon just to produce a random result. Why not hand the whole business over to a computer? If we accept the ends of a particular process as legitimate then it is hard to resist any technique which increases the efficiency by which it is achieved. The car *is* faster than the horse-drawn coach, and it is hard to argue for the latter if you want a quick means of transport. Insofar as the development of technology is the process by which any given task or activity or service is done more efficiently, then technology wins for itself (and those advocating it) political authority and legitimacy.

So technology seems to provide winning solutions to political problems. But its impact does not end there. Technology could also be said to be contagious – it spreads rapidly and widely. It is not just employed to develop new forms of existing techniques; it is also applied to new areas of life. Thus automation technology is not just applied to the factory floor but also to the office, in the form of the word processor, electronic mail, and so forth. More and more areas of life are coming to be organised around technology. This kind of spread is not gradual; it is like a nuclear chain reaction. One neutron hits the nucleus of another atom; this releases two neutrons which strike two further nuclei, each of which yield two neutrons of their own, and so on. The spread is very rapid. So it is with technology. It does not spread at a steady pace; the rate of applications increases with each year so that not only do we have new technology but also new places for it to be applied. (To comprehend the speed of spread, consider this game. Imagine you are offered two alternatives. Either you can accept one dollar on day one, two dollars on day two, three dollars on day three, and so on until day 30. Or you can have one cent on day one, two cents on day two, four cents on day three, eight cents on day four, and so on until day 30. Which would you accept? If you chose the first option you would be 465 dollars better off; but if you chose the second option, you would be a millionaire. The spread of autonomous technology is like the latter.)

The combination of technology's self-legitimating rationale and its proliferating form give it the appearance of autonomy. It cannot be resisted both because the arguments against it seem weak and because the attack from it comes on so many fronts. But we must not leave the argument here. We need to recognise that the force of this interpretation rests on a set of judgements about technology and technological rationality.

The third and final stage in the argument is the ethical or political claim made about the values embodied in technical change. According to the autonomous technology school, the spread of technological rationality within society severely diminishes the quality of life for people. It is supposed that individuals confront two types of decision: the moral and the instrumental. The first is about what ought to be done; the second is about how it can be achieved. The latter is served by technological rationality, the former by morality. Morality, according to this argument, provides the over-riding purpose for life; it establishes human goals. Instrumental reason is of secondary importance. The autonomous technology critique claims, however, that this arrangement has been reversed. The dominant question is no longer about what we ought to do; it is only about what we can do.

In illustrating this process, writers focus on activities such as art or sport. They contend that the higher values of human struggle and expression have been lost to a technocratic order. Sport is now a competition between pharmaceutical companies as much as athletes; music is about the mass production and mass consumption of a 'life-style'. Returning to the example of Alan Turing, the autonomous technologist might conclude that Turing's treatment was proof of technology-out-of-control. First, 'homosexuality' was defined as a technical problem. Having been identified in this way it is inevitable that a technical solution is sought. All other solutions or approaches lack the conviction of the technological one. And arguably the same thinking continues in the new guise of psychotherapy.

In summary, the theory of autonomous technology claims that the form and character of society, including its politics, are dictated by a technological rationality. It is this which dictates the outcome in any debate over alternatives. Technology's 'authority' is supplemented by its spread into all areas of life. The effect of this is to reduce human agency and choice, to eliminate debate, and to render individuals powerless. In 1941, Marcuse anticipated a world in which 'business, technics, human needs and nature are welded together into one rational and expedient mechanism'. Under these conditions,

> all protest is senseless, and the individual who would insist on freedom would become a crank. There is no personal escape from the apparatus which has mechanized and standardized the world. It is a rational apparatus, combining utmost expediency with utmost convenience, saving time and energy, removing waste, adapting all means to the end, anticipating consequences, sustaining calculability and security (Marcuse, 1941, p. 419).

But is such an account coherent, is it accurate?

There are many problems with the theory of autonomous technology. Not least is its failure to account for differences in the form, organisation and effect of technology in different political settings. Broadcasting systems, for example, vary from country to country. Autonomous technology also tends to exaggerate the 'rationality' of technology, investing it with a perfection that it evidently lacks. We need only think of technological disasters and pollution to be made aware of technology's fallibility. Finally, in attributing all change to technology, the theorists of autonomous technology remove all responsibility from those who presently wield power or those who seek to wrest it from them. The idea of autonomous technology can be used as a disguise for culpability. During the Vietnam War, military officials blamed the illegal bombing of Cambodia on computers (those 'damned things') to which they claimed they were 'slaves' (Burnham, 1983). In fact, the illicit targets had been programmed into the computers by the military.

For each or all of these reasons, the theory of autonomous technology may be vulnerable, and as a result a more moderate account of technology's determining role may be adopted.

Technological determinism

The theory of technological determinism bears some resemblance to that of autonomous technology. They both represent technology as the driving force of social change, but they differ in their portrait of how this process works. The theory of technological determinism makes no particular claims about the ideological rationale provided by technology or about the extent of its impact. It does, however, contend that technology sets the conditions for the operation of the political system, including the political agenda, even if it does not determine the policy output. Within this general approach, there are two strands to technological determinism. One refers to the tendency of technology to force change on society; the second is directed specifically at the type of change involved.

Technical change can appear to present people with no choice; it constitutes a demand to adapt. In this respect, technological determinism resembles its autonomous cousin. So a 'soft determinist' like de Sola Pool writes of the emergence of communications technology as having 'profound effects on civilisation' (Pool, 1990, pp. 7–8). It is a process that cannot, ultimately, be resisted. Governments may resist the new technology, but 'those countries that are restrictive of innovation (and they will probably be in a majority) will lose out in competition to those countries that take full advantage of new possibilities' (Pool, 1990, p. 148). Defending this idea, Christopher Freeman writes: '"new technological systems" can offer such great technical and economic advantages in a wide range of industries and services that their adoption becomes a necessity in any economy exposed to competitive economic, social, political and military pressures' (Freeman, 1987, p. 5). In this guise, technological determinism preempts choice.

The second aspect of determinism focuses less on choice and more on the surrounding political process. The political order, it is contended, has to be explained in terms of the technology upon which a society relies. Technological determinism takes literally Marx's claim from *A Contribution to the Critique of Political Economy* that 'The mode of production of material life conditions the general process of social, political and intellectual life' (Marx, 1975, p. 424). The sort of processes involved can be read into this extract from *Capital*:

> Modern Industry never looks upon and treats the existing form of process as final. The technical basis of that industry is therefore revolutionary, while all earlier modes of production were essentially conservative. By means of machinery, chemical processes and other methods, it is continually causing changes not only in the technical basis of production, but also in the functions of the labourer, and in the social combinations of the labour-process. At the same time, it thereby also revolutionises the division of labour within the society, and incessantly launches masses of people from one branch of production to another (Marx, 1954, p. 457).

What Marx appears to be saying is that technology establishes a particular set of power relations.

The economist Robert Heilbroner (1972) makes a spirited attempt to defend this 'strong determinist' reading of Marx. He presents his argument in two stages. In the first, he seeks to establish that technological change follows a pre-ordained path, and is therefore not subject to political or social influence. The second stage then addresses the question of whether technology shapes society.

He defends the first proposition, that technology acts independently, on three grounds:

1. If the same discovery is made simultaneously in different contexts, this suggests that technology contains a logic of its own which is independent of its associated agents.
2. The tendency of technology to develop in steady stages, rather than in sudden leaps, suggests a process of evolution contained within the logic of technology.
3. There is an element of predictability to technology's development. Again this suggests that technology contains an inner logic.

If these three propositions hold, says Heilbroner, then we have grounds for seeing technical change as an independent process.

He then moves to the second stage of the argument, that technical change determines social change. Here Heilbroner points to two features of society which seem to be determined by technology:

1. The social composition of the labour force – the level of education, skills and so forth – seem to correlate directly with the 'needs' of the technological base;
2. The hierarchical organization of work is also affected by the technology.

Heilbroner observes that 'different technological apparatuses not only require different labour forces but different orders of supervision' (Heilbroner, 1972, p. 34).

At this point, though, Heilbroner backs down from the attempt to defend 'strong determinism'. While it might suffice as a possible description of a particular moment in capitalism's development, it does not provide a general theory. Strong determinism allows no place for the social processes contained within technical development, nor for the influence of context on the development of technology. As Hughes (1983) showed in his history of electricity, both the process of innovation and implementation are influenced by local and political factors. Heilbroner opts instead for a soft determinism, which requires that technology only sets the conditions for politics, rather than ordering the form of political structures. This modified form of determinism has received a fuller exposition in the works of the economist J.K. Galbraith and the sociologist Daniel Bell.

Galbraith works with the notion of 'technological imperatives'. He argues that modern technology demands the creation of a particular type of political state. Without it, technology will not function, and without technology no wealth can be generated; without wealth, political demands cannot be met and political legitimacy cannot be maintained. The development of technology, argues Galbraith, is essentially the progressive application of the division of labour. He writes:

> Technology means the systematic application of scientific or other organized knowledge to practical tasks. Its most important consequence, at least for purposes of economics, is in forcing the division and subdivision of any such task into its component parts (Galbraith, 1974, p. 31).

Each technical advance depends on separating out the key elements in any process and then applying scientific or technical knowledge to each. The production line is the classic example of this approach.

The development of technology has a profound effect on the role and function of the state. The state has to respond in particular ways, as if it were being prompted by technology. The first consequence of the increasing division of labour is the need for central coordination. Without some mechanism for collating information and for ensuring the compatibility of components, the division of labour will collapse into chaos. But coordination at the centre is not the only problem posed by the development of technology. Given that this development depends on the application of scientific research, such research has itself to be funded and organised. The responsibility for this is unlikely to fall on the manufacturers because with the increasing division of labour the knowledge being applied becomes ever more abstract, and its particular usefulness more uncertain. It does not make sense for individual entrepreneurs to bear the cost of this research. Those burdens have to borne by a central source of funds (the state) and the institutions of higher education which it subsidises. Similarly, the products of the new technology have to be assured both of a demand and of an infrastructure that can accommodate them. Cars need roads and licensing authorities, and much else besides. In summary, Galbraith argues that the particular character of technology establishes the need for a state which can coordinate, fund and organise the process by which the division of labour proceeds. The technology calls the state into being; it determines the need for, and form of, the key political actor.

A parallel argument to Galbraith's can be found in Daniel Bell's *The Coming of Post-Industrial Society* (1973). He, like Galbraith, sees technology as setting the parameters to the political order and sees planning as a vital feature of political practice. But where Galbraith focuses on the political structure, Bell concentrates on the personnel who are to inherit this order and on the kind of decisions they have to make. For Bell, the emergence of new technology, based on the application of science to knowledge and information, creates a new elite who are to take responsibility for the emerging political order. Bell summarises his main argument as follows:

> The rise of the new elites based on skill derived from the simple fact that knowledge and planning – military planning, economic planning, social planning – have become the basic requisites for all organised action in a modern society. The members of this new technocratic elite, with their new techniques of decision-making (systems analysis, linear programming, and program budgeting), have now become essential to the formation and analysis of decisions on which political judgements have to be made if not to the wielding of power (Bell, 1973, p. 362).

The members of the new elite are responsible for establishing the political values which will determine the processes at work within the post-industrial society (ibid., p. 337). The elite will have a crucial role. As he comments, 'the central point about . . . the post-industrial society . . . is that it will require more societal guidance, more expertise' (ibid., p. 263). The rise to power of the new elite is determined by the replacement of the market by planning as the organisational form of the new society, a change that is itself determined by the revolution in the technological base.

Bell, again like Galbraith, does not see the determinism generated by technological change extending to the actual character of political decisions. He does, though, see the agenda and those who decide upon it as being the creatures of developments in technology. In this sense, both Bell and Galbraith present a soft version of technological determinism. We can illustrate their model as follows:

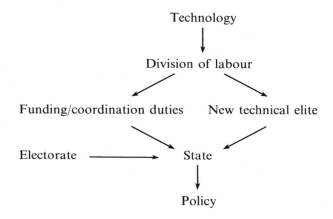

Political processes, rather than political outputs, are determined by the character of society's technological base. The base establishes the need for a particular division of labour and the accompanying political order to organise it.

Assessment of the theories of autonomous technology and technological determinism

The specific features of the division of labour and the demands it puts on the state need not detain us here. Our concern is with the general structure of the argument whereby technology is seen as the author of orders to which the political structure has to bend. While the effect of autonomous technology is measured ideologically, in the way people reason and make choices, the effects of technological determinism are confined to the political structure. Technological determinism says little about the choices made; its concern is more with how they are to be realised. A number of objections have been aimed at the idea that technology can be said to determine social structures, whether the argument takes the form of autonomous technology or technological determinism. One problem stems from the vagueness of the theories. It is noticeable that a wide range of verbs are used when describing the impact of technology on politics. Sometimes the word 'determine' is used, suggesting a fixed causal link, at other times words like 'shape' or 'guide' or 'influence' are employed, suggesting a less certain connection. 'The concept of determinism', writes Winner (1986, p. 10), 'is much too strong, far too sweeping in its implications to provide an adequate theory. It does little justice to the genuine choices that arise, in both principle and practice, in the course of technical and social transformation'.

Winner focuses this confusion by drawing attention to two ways in which determinism might work. He begins with the simple observation that, almost by definition, changes of technology entail changes in the social world. Technology always changes what people can do and therefore how their lives are organised. The point is to establish the character of the change involved. Winner distinguishes between being 'determined' and being 'conditioned'. The first refers to the idea that certain behaviour or institutions are *caused* by technology – technology acts as the driving force. Being conditioned, on the other hand, suggests a form of 'reverse adaptation': 'the adjustment of human ends to match the character of available means' (Winner, 1977, pp. 83, 229). The latter entails a much less malign and less powerful role for technology. People adapt their goals to fit their technical environment; they retain substantial choice and relative autonomy, and are not slaves to their technology. The first form of determinism, however, suggests that no such control is available; people are merely creatures of their technology. The question is whether this constitutes a coherent claim.

How does an inanimate object – technology – determine political processes? We may say that a wall 'blocks' our path or gravity 'denies' us the capacity to fly unaided. But while such figures of speech contain important truths (we cannot walk through walls), they only tell half the story. The barriers set by walls or gravity are sometimes, as in the case of the former, a result of human decisions which themselves need to be accounted for. And though gravity is not a human construction, we have to account for the desire to fly. Some things become barriers because of the actions we have chosen. In other words, it is wrong to speak of determinism here, except in the weak sense of reverse adaptation.

Nathan Rosenberg (1981) attacks the stronger form of technological determinism by arguing that technology does not 'determine' anything because its own role is itself the result of economic interests which call certain technologies into existence and which establish the relations around them. Technological determinism looks at the problem from the wrong end. It is not technology that determines political practices, but politics that determines the use of technology. This view is echoed by Goldhaber, who sees technical progress as a purely political phenomenon: 'technological progress is always guided by values and interests that come from outside technology' (Goldhaber, 1986, p. 14). The dominance of politics makes more sense, for example, of the development of space technology. Robert Goddard offered his pioneering work on rocket technology to the US government during the Second World War. His offer was rebuffed.

Stephen Marglin's (1978) account of the rise of the factory offers similar evidence to counter the idea of technological determinism. According to the determinist account, mass production techniques should predate the division of labour (workers and managers, skilled and unskilled) that characterised the factory system. But Marglin shows that, in fact, the factory form of organisation preceded the technology. Marglin's explanation is that the factory was not determined by technology but devised by capitalists. Their economic interests were served by the creation of that particular form of organisation. The organisation and implementation of technology was politically determined, rather than the reverse.

In criticising the determinist models we have moved to a view which emphasises the central role played by political choice in the development of technology. But is this view any more coherent?

Political choice

While there is a clear resemblance between the theories of autonomous technology and technological determinism, they share very little with the political choice model. Here the relationship between technology and politics is completely reversed. Where the previous two theories put the explanatory burden on technology, here political choice accounts for the form of the technology. Political choice is a case of mind over matter; the others are matter over mind. It can be illustrated simply as:

Political interests/values   ⟶   Technology

The political choice model can be seen at work in the claim that the development of technology follows directly from the fulfilment of human needs. It is those needs that create the reason for developing the technology:

> Humans have a need for water, so they dig wells, dam rivers and streams, and develop hydraulic technology. They need shelter and defense so they build houses, forts, cities, and military machines. They need food, so they domesticate plants and animals. They need to move through the environment with ease, so they invent ships, chariots, carts, carriages, bicycles, automobiles, airplanes, and spacecraft (Basalla, 1988, p. 6).

This form of argument does not have to be couched in terms of general human needs. It can equally apply to a particular political structure of needs and/or interests. 'Class needs', for example, can do the work of 'human needs'. One version of this, often associated with Lenin, makes the technology entirely passive and neutral. Following the 1917 Russian Revolution and a brief flirtation with the idea of workers' control, Lenin adopted a form of industrial management (together with mass production) which mirrored that adopted in the West (Brinton, 1972). For Lenin, what mattered was who used the technology, not what shape it took. In itself, the technology was apolitical; it was simply the most sophisticated method of industrial production. Its application in the Soviet Union would maximise the availability of goods to be used for the benefit of all. Lenin accepted the argument that the development of technology was measured only by increasing levels of efficiency. Technical change was merely the application of scientific advance of the kind which Marx imagined would lead to the emancipation of the working class. Technology itself was irrelevant; what mattered was who owned it and whose interests they served.

Lenin's argument now finds little favour among the left, although it formed a constant theme in left-wing thinking for much of the first part of the twentieth century (Street, 1981). Its best-known critic has been Harry Braverman. In his *Labor and Monopoly Capital* (1974), Braverman argues that technology itself cannot be regarded as independent of political or class interests. Science and technology, says Braverman, are themselves ordered according to capitalist interests. The driving force behind the funding of science and the development of technology is the reduction of labour costs and the workers' control over their work. Albury and Schwartz summarise the general argument: 'the shape of particular disciplines and specialities in science and technology has been moulded by the determination of the capitalist class to mobilize all resources in their attempt to maintain ownership, power, and control' (Albury and Schwartz, 1982, p. 67). The word processor, for example, is not to be seen as a technical improvement on the typewriter, but rather as a way of rationalising the office and disciplining office workers.

Feminist theorists have also adopted this line of argument to explain the way in which both the design and the form of much technology is intended to ensure the maintenance of male domination (Cockburn, 1983; Barker and Downing, 1980). In this respect, their arguments echo those of the political choice model. They see political interests as determining technical form.

Where the autonomous technologists want us to read our society in the language of its technology; the political choice model wants us to read our technology in the language of our society. A divided society makes technology which reproduces – and reinforces – those divisions. Political interests determine the organisation of production and the use to which it is put. Technology is value-laden. Such an approach is used to explain why, for example, car manufacturers have opted for relatively cheap, but less effective, seat belt technology in preference to more expensive, but safer, air-bag systems (Reppy, 1979). The car embodies a set of priorities: profit takes precedence over safety; manufacturers take precedence over drivers and passengers.

Assessing the political choice model

Recalling the distinction we drew earlier between internal and external portraits of technical change, it is clear that the political choice model is concerned only with giving an external account. It shows technology as responding to the commands of the political realm, however that is structured. There are, though, two causes for concern about this version of the politics–technology link. The first is that by concentrating on the external influences on technical change, it allows no room for the internal dynamics of technical innovation and development. The relations and interests which combined to actually produce computers, for instance, are subsumed within a set of wider political forces. The political choice overlooks the micropolitics of innovation, and fails to explain how the larger political interests are realised in the application of scientific knowledge.

A second line of criticism flows from the first. The political choice model imbues the political structure with the capacity to control and anticipate change. In doing so, the model makes little allowance for the unintended consequences of technology (pollution, accidents, and so on) or for the systemic character of technology which is expressed in the idea of technological momentum (Hughes, 1983). In such circumstances, it makes more sense to talk of political interests *coping* with technology rather than controlling it.

UNTANGLING THE WEB

So far we have seen the bald outlines of the competing accounts of the link between politics and technology. In this final section I want to see whether, rather than rejecting any or all of the above theories, we can defend a more eclectic approach which draws on the strength of each. This approach is characterised by Jamison (1989, p. 533) as one which seeks 'to link the "technological imperative" with the cultural context, the internal understanding of technology's development with the external application of social needs and desires'. Such an attempt is born of consideration that the general theories are too indiscriminate in their approach both to technology and politics. While, for example, the theory of autonomous technology may account for the general trend in weapons development, it does not discriminate between the way different political systems organise their weapons industry.

The very idea that we can generate a complete theory of political and technical change seems misconceived. There is too much to be accommodated. As W. G. Runciman (1989, p. 336) points out, an immensely complex set of cultural and structural conditions have combined with the new industrial technology to produce 'industrial capitalism'. In fact, the processes of social and technical development are characterised by a high degree of unpredictability, with consequences rarely matching intentions. The emergence of industrial capitalism has, Runciman writes (ibid., p. 340), 'to be treated as random not only in the sense in which all mutations and recombination of practices have to be treated as random but in two further senses: first, the juxtaposition of the several initial conditions which were jointly necessary was entirely fortuitous; and second, since the transformation of the mode of production was the first of its kind, there was no goal which the agents who created it could have in view to aim at'. To see change in this way is not to discount the value of theory. Instead, it argues for theories which acknowledge the limits to generalisation and the contingency of the conditions in which change occurs. The links between politics and technology are in constant flux:

Changes in political conditions ⟷ Developments in science/technology

New political demands/need ⟷ New technical possibilities

We need different types of account when we are explaining state control of technology than when we want to see how particular technologies constrain individual choices. This kind of synthesis is proposed by Thompson (1989, p. 140), who argues for the need to combine 'the actions of the state', the particular character of the technologies involved and their context, and 'the role played by key actors'. A similar plea is made by Rudig (1990) when he urges the need to combine the sociology of technological development with the politics of the policy process.

A model of the approach I have in mind can be found in William McNeill's *The Pursuit of Power* (1983), in which he tells the combined story of changing weapons technology, military strategy and government policy from ancient times to the present day. McNeill's account casts doubt on any over-simplified version of technological determinism. While technical innovations do alter the conditions under which war is fought and military forces are organised, there is no straightforward pattern of determination. This is because the development of military technology is itself dependent both on political priorities and subsidies. There would be no need for weapons without the will to wage war. But the story does not end there. Political will on its own is not sufficient. States vary in their capacity to organise weapons production and to establish an efficient military structure. Describing political decisions in Britain in the early twentieth century, McNeill analyses a trend towards ever-increasing naval expenditure. It was, he says,

fuelled by a runaway technological revolution as well as by
international rivalries and the changed structure of Great Britain's domestic politics. A powerful feedback loop established
itself, for technological transformations could not have procee-
ded nearly so rapidly if economic interest groups favouring
enlarged public expenditure had not come into existence to
facilitate the passage of bigger and bigger naval bills. Each naval
building program, in turn, opened the path for further technological change, making older ships obsolete, and requiring still
larger appropriations for the next round of building (ibid., pp.
277–8).

This notion of a 'feedback loop' is central to understanding this
version of the politics–technology connection.

The Pursuit of Power shows that technology is developed and
applied in accordance with political criteria. Politics plays a role
both in terms of the values and judgements at play, and also in the
institutional structures and political interests that transmit and
organise these ideas. At the same time, McNeill's evidence suggests
that politics is not autonomous. Changes in technology, which can
neither be predicted nor controlled, have an impact on the
conditions which frame the political arena. The competition
intrinsic to war or hostility is enough to demonstrate this. Writing
of the nineteenth century, McNeill says: 'Each French breakthrough
[in war technology] provoked immediate countermoves in Great
Britain' (ibid., p. 227).

The relationship between technology and politics is, in other
words, only partly forged by changes in technology. This point is
reinforced by George Basalla in his book *The Evolution of
Technology* (1989). Basalla, in arguing against the idea that human
need determines the course of technical change, cites the case of the
car:

A search for the origins of the gasoline-engine-powered motorcar
reveals that it was not necessity that inspired its inventors to
complete their task. The automobile was not developed in
response to some grave international horse crisis or horse shortage. National leaders, influential thinkers, and editorial writers
were not calling for the replacement of the horse, nor were
ordinary citizens anxiously hoping that some inventors would

soon fill a serious societal and personal need for motor transportation. . . . In other words, the *invention* of vehicles powered by internal combustion engines gave birth to the *necessity* of motor transportation (ibid., pp. 6–7; his emphasis).

At the same time the ability to develop a technology is a function of both the way the political process determines what priorities are to operate and the capacity of that system to respond to technical change. Technology can appear to be autonomous, not because it is changing independently, but because the political system fails to control it. This is the argument that can be detected in McNeill's account of pre-First World War Britain: 'it seems correct to say that technical questions got out of control on the eve of World War I in the sense that established ways of handling them no longer assured reasonably rational or practically satisfactory choices. Secrecy obstructed wisdom; so did clique rivalries and suspicion of self-seeking' (McNeill, 1983, p. 298). The political process helped to 'create' the determinism, in that it made possible certain results.

A similar conclusion emerges from Hughes' discussion of the emergence of national electricity generation in Britain. He draws attention to the way 'technological momentum' and 'political structures' are linked, albeit not in a general nor in a consistent way. They cannot be separated, even when appearances suggest otherwise. It may have seemed that in Britain the 'high-momentum [electricity supply] systems of the interwar years gave the appearance of autonomous technology'. But, Hughes goes on, these impressions were deceptive: 'In fact, the British style accorded nicely with prevailing British political values and the regulatory legislation that expressed them'. For Hughes, the appropriate analogy for technically-influenced change is 'momentum': 'Momentum does not contradict the doctrine of social construction of technology, and it does not support the erroneous belief in technological determinism. The metaphor encompasses both structural factors and contingent events' (Hughes, 1987, pp. 79–80).

The political impact of technology is not just a matter of the structure of the political process. As Hughes writes (ibid., p. 2), 'electric power systems, like so much other technology, are both causes and effects of social change'. A similar relationship is evoked by McNeill when he concedes that the lack of control over technical decisions in the run-up to 1914 was itself partly caused by the nature of the technology: 'the mathematical complexity of the problem – a complexity which clearly surpassed the comprehension of many of the men most intimately concerned – deprived policy of even residual rationality' (McNeill, 1983, p. 298).

No two technologies are exactly alike in the way they control or can be controlled. 'Technologies vary', writes Collingridge (1983, p. 230), 'in the ease by which they may be controlled through the normal policy process'. Some technologies require special types of political control. Winner describes the atom bomb as just such a technology. It

> is an inherently political artifact. As long as it exists at all, its lethal properties demand that it be controlled by a centralized, rigidly hierarchical chain of command closed to all influences that might make its workings unpredictable. The internal social system of the bomb must be authoritarian; there is no other way (Winner, 1985, pp. 32–3).

While Winner acknowledges that the bomb is a special case, he claims that certain technologies do seem to 'require' certain accompanying political structures. But he raises the important question, not recognised by those who stress technological autonomy or determinism, or by those who concentrate only on the internal life of technology, that the link between a technology and a political structure may not be completely rigid. There may, in short, be other ways both of accommodating existing technology or of redesigning it. And for political scientists, the question then becomes why particular possible combinations were arrived at. It is no longer a matter of simply explaining either the technology in terms of the politics, or the politics in terms of the technology.

CONCLUSION

The main concern of this chapter has been to ask how technical and political change might be linked. We began by examining the general features of technical change, the difference between innovation and implementation, between internal and external accounts of change. This discussion drew attention to the network of interests which combine to generate technical change. It established the importance of political institutions and values to such change, but it said nothing about cause and effect.

Three possible models of the politics–technology relationship were considered – autonomous technology, technological determinism, political choice. Because each was pitched at a high level of generality, they tended to miss the detailed processes by which politics and technology interact.

My central claim here is that there is no single approach to the relationship between politics and technology. In analysing the connection we need to retain a spirit of eclecticism in which we recognise differences in both technical and political form, and where the relationship we identify will depend on each. This conclusion is not meant as an unqualified embrace of empiricism nor as a renunciation of theory (were either possible). It is rather a rejection of a general theory, and a request for a broader range of hypotheses about how technology and politics may combine.

It is with these thoughts in mind that we now turn our attention to a particular element in this complex web: the state. It is one thing to observe that the form of technology and of political structure will affect the degree of control involved; it is quite another to see where different technologies and different forms of political organisation intersect.

3 The State and Technology

INTRODUCTION

We may not have to accept the technological determinist's suggestion that the modern state has been transformed by technology, but we cannot, however, avoid the conclusion that the state has found itself intimately bound up with technology. The state can be found promoting the introduction of technology, using it to sustain national prestige, deploying it to maintain internal and external security, and subsidising its research and development. As Donald MacKenzie observes, 'Take away the institutional structures that support technological change of a particular sort, and it ceases to seem "natural" – indeed it ceases altogether' (MacKenzie, 1991, p. 384). The state is one such institutional structure. Political leaders make a point of attending the unveiling of new technologies. 'Large scale technology initiatives', remarks Peterson (1989, p. 12), 'can often pay considerable and immediate political dividends'. Nations compete with each other to be the first with a new technological breakthrough. Wernher von Braun, one of the USA's rocket engineers, once wrote of the decision to put men on the Moon: 'A country from which a major proportion of mankind expects dynamic leadership has no choice but of either taking up the historical challenge of the day or of stepping down from the position to which fate has lifted it' (quoted in Marsh, 1985, p. 9).

A state does not get involved with technology just for its prestige value. Surveillance technology and military technology are crucial elements of a modern state's armoury, part of its survival mechanism. Equally, without the contributions of government, many technical endeavours would be stillborn. And without state regulation, many technologies would be unusable or prohibitively dangerous. So it goes on. The state, for good or bad, finds itself playing a variety of different roles in the development, deployment and management of technology.

46

This relationship between the state and technology, argues Roger Williams (1990b, p. 217) has prompted the emergence of a 'new politics' based on science and technology. In this new order governments depend for their survival on their ability to direct the development of technology. Political argument centres on questions of how best to promote innovation in, and implementation of, new technology. Political interests are shaped in response to the dissemination and development of technology.

One of the central questions prompted by this new alliance is the degree to which governments are able to control the course taken by technology. Does a government *direct* the development of technology? Or is it forced merely to respond to the demands made upon it by technology? What sort of pressures, values and judgements are at work in the state's response to technology? Not surprisingly, there are no easy answers because there is no single dimension to the problem. States differ in their structures and in the principles that underlie them; countries differ in the environment in which they exist; and technologies differ in the problems they pose at their various stages of development.

That most familiar feature of modern industrial life, the car, vividly illustrates the range of issues that are confronted in either exercising or analysing political control over technology. First, it is hard to establish where the car began – with the invention of the internal combustion engine, or with Boyle's law and the link between pressure and volume, or with Henry Ford and mass production? Without a clear starting point, there is no easy route to establishing responsibility for its development. Equally, the subsequent effects of the car generate an almost endless series of issues with which governments have to deal. Finally, confronted with the car and its impact, what choice does a government face but to legitimate and subsidise its development and use? It is no longer in a position to accept or reject the car.

But while the car seems to pose a general set of problems with which all governments have to deal, we must also notice that governments do not respond in an identical fashion. There are variations in forms of traffic management, safety regulation, pollution control, and so on. These differences may have a number of explanations. There may be the context in which the decision is taken. The car poses a different problem to the government of a developed country than it does to the government of a less developed country. On the other hand, the governments of two developed countries do not respond in the same way to the problem of the car.

The source of such variation is the concern of this chapter. It will enable us to say something about the relationship between states and technology, about how and why states seek to control technology. There are three dimensions to the state–technology relationship:

1. Types of involvement.
2. Political structures of control.
3. Limits of control.

These form the structure of this chapter.

TYPES OF STATE INVOLVEMENT

It is already apparent that there are many different ways in which the state can become involved in the development of technology. It may, as in the case of nuclear power, play a dominant role throughout the history of the technology. Alternatively, the state's involvement may come late and have limited impact. At whatever stage the state becomes involved, its motives for doing so may take a variety of forms. At one extreme is involvement inspired by the desire for kudos; the political elite seeking to associate itself with a technological breakthrough to which it may have contributed virtually nothing, but through which it hopes to enhance its political image (scientists and technology sometimes appear on stamps and banknotes). The relationship is like that between a parasite and its host. The corollary of such involvement is the desire to avoid association with a technological failure – whether or not there has been any previous state involvement.

State involvement may, though, represent more than simple exploitation of potential political pay-offs. The state may see a technology as a way of serving certain practical political goals. Both the technology of surveillance and of defence would fit this category. Here the state acts as a customer. Another form of direct state involvement is the way the state seeks to regulate the building or operation of technologies. From the imposition of safety guidelines for children's playgrounds and the granting of planning permission for new buildings, to the rules governing the production of pharmaceuticals or the operation of the airline industry, the state is involved in the regulation of technology. A final role may involve the state as an underwriter, providing the resources or support out of which technology emerges. State funding has played a vital part in the development of weapons technology, microelectronics and biotechnology. In such circumstances the state is not necessarily the direct beneficiary or user of the product, although its interests may be served by the results.

These three roles – the state as customer, regulator and underwriter – are discussed in more detail below. My aim is to expand upon the details of each form of involvement and draw out some of the implications of these.

The state as regulator

The technology of mass communications emerged at roughly the same time in both Britain and the USA. Furthermore, that technology took roughly the same form. But despite these similarities, each country evolved quite different broadcasting formats. In the USA, the system was largely unregulated and commercially driven. In Britain, there was a centrally organised network, governed by a public service ethos. Though the basic building blocks were similar, the resultant technological systems were quite different. They differed in the forms of access and control they allowed and in the kind of service they provided. The two systems differed markedly, for example, in how they treated their listeners. The commercial model tended to treat its audience as consumers, making them into markets for their advertisers' products; whereas the public service model created an audience of 'rational citizens' who could be defined as 'the public' (Lewis and Booth, 1989, pp. 6–8). There is no *technical* explanation for the contrast; it lies in the political criteria which organised broadcasters. Crucial determinants of these criteria were the respective governments and the considerations they brought to bear upon the development of broadcasting technology.

The differences in broadcasting systems reflected very broad assumptions about the role and responsibilities of government. The federalism of the United States made the argument for limited control, while Britain's centralised power structure argued for monopoly control. There are, however, more precise reasons for the development of the different systems. The emergence of radio was also framed by the interests that surrounded other forms of communication, most obviously telegraph and telephones. These had their own institutional and commercial interests which were threatened by the new technology. Most important for the development of radio were the military (and hence security) interests which came with communications technology. The development of wireless telegraphy was crucially linked to the navy. Lewis and Booth write: 'It was the use of wireless telegraphy at sea as a means of saving lives and controlling the movements of shipping, especially of naval fleets, that formed the general conception of the medium' (Lewis and Booth, 1989, p. 16).

The form radio took was not just shaped by security associations. Radio acquired commercial interests. These were registered in the way the new technology was patented; that is, the way in which the rights to profit from, and to use, the technology were determined. The allocation of patents was extremely complicated because of the many components that went towards the broadcasting system (ibid., p. 17). In Europe, however, two companies (Marconi and Telefunken) were able to use their patents to negotiate very restrictive deals with European governments. This tied the hands of the governments and created conditions favourable to the formation of broadcasting monopolies. Suspicion of these developments in Europe led the US government to open up the market (ibid., p. 21). Government action, in other words, was an important component in explaining variations in the development of radio.

Apart from the control of patents which related primarily to the hardware of radio, another vital issue was the distribution of the airwaves. Radio frequencies do not constitute an unlimited resource, and access to them has to be administered. The question is how? In the US, the presiding policy is one which 'seeks wherever possible to fill frequency space rather than find reasons to deny its use'; in Britain, by contrast, 'the onus is on the citizen to show cause why s/he should use the frequency spectrum at all' (ibid., p. 22). Each approach has quite different results for the kind of voices and sounds that the listener hears. Lewis and Booth explain this divergence in two ways. Firstly, the US government was subject to commercial push from radio networks which had emerged in the post-First World War period. Given its disposition against federal control, the government favoured a 'hands-off' approach which allowed the airwaves to be filled. Secondly, such a policy accorded with the constitutional authority given to freedom of expression and information. In Britain, by contrast, there was a general structure of central control reinforced by highly restrictive secrecy legislation. When these institutional factors are combined with a situation in which military interests are given priority in the distribution of frequencies, it is not surprising that Britain's version of broadcasting technology took on a very different character to that in the US (ibid., pp. 22–3).

Hughes (1983) tells a similar story in his account of the development of electrification. He observes how different political structures and processes proved more or less hospitable to the introduction of electricity. He writes of the resistance to electrification in England, explaining it in terms of

> the pervasively conservative attitudes of economic, political and technological interest groups that were content with their status and future in a world without electrical power systems. For example, electric power supply threatened to overrun and weaken the authority of local governments, to displace the established gaslight technology, and to devalue investments in that older technology and its institutions (Hughes, 1983, p. 66).

In short, the fortunes of the technology were largely determined by the political interests around it.

Were we to consider only governments' roles in the regulation of broadcasting or electrification, we might reach the simple conclusion that in the US technology is given a free hand, whereas in Britain it is closely monitored and controlled. Such a conclusion is, of course, too easy. We can find examples of technologies in which the US imposes much more restrictive forms of regulation on their operation than occurs in Britain. Congress has imposed restrictions on the movement and sale of US computers, extending this ban to areas formally outside its jurisdiction. British computer firms have been prevented from exporting US hardware. In the medical field, the US has banned the use of drugs which have remained available in the UK. The building of the Channel Tunnel by French and British companies revealed a similar contrast. Many more British than French workers died in the course of the tunnel's construction. Although it is impossible, at this stage, to account fully for this disparity, one factor may be the different safety standards imposed and achieved by the respective governments. In this sense, the state as regulator has an impact on the operation of technology.

The state as customer

A government's relationship with technology is not, though, confined to that of regulation (where the state is at one remove). Technology can be an integral part of the state's activities. Military weaponry is the most obvious, and most frequently cited, example. Because of this, and because it is considered elsewhere in this chapter, I want to focus on some marginally less familiar instances of the state as customer.

Computer technology plays an ever-increasing role in the conduct of governments. One of the prime purposes of the computer is to aid in the collation and management of information. Governments use computers to store data on their citizens and to help administer services for them. Information technology has been integrated into the administration of the state. Indeed, it is conceivable that, without the introduction of computer technology, changes in the administration of, say, a complex system of welfare benefits would be impossible. Computer technology is also employed in the pursuit of the state's security role. Not only is the data collected for tax or other purposes used in the investigation of criminal activities, computer technology is also deployed in direct forms of surveillance (Campbell and Connor, 1986).

The state's use of technology is most obviously demonstrated by the police. Police use of technology is not confined to the high technology of the computer. The more mundane technologies of the car and radio communications have transformed policing, as have the technologies of public order – CS gas, water cannons, plastic bullets, riot shields, and so on (Stephens, 1988, pp. 60–1, 79–86; BSSRS, 1985). Few of these technologies were exclusively developed for the police, nor was their introduction an inevitable consequence of their invention. We have only to compare the different technologies employed by police forces around the world. What matters is the decision of individual states – albeit under pressure from manufacturers, police associations, and other interested parties – to acquire the technology.

The state is not, it is true, an ordinary customer. It commands crucial markets and vast resources, and as such it can have a major impact upon the design of the technology. But equally, as with any customer, it can be persuaded to adopt technology 'off the shelf'. Nuclear power is an example of the former; while computer acquisition is an instance of the latter. Whatever is the case, the state plays an important role in purchasing and introducing technology, and this can effect both the state's operation and the character of available technologies.

The state as underwriter

In 1979, the British government decided to create a private company, Inmos, to act as an indigenous source of microchips. This required buying in the necessary talent from the private sector (and had the incidental effect of creating two individual millionaires). Inmos was subsequently bought up by the private sector. In the early 1980s, the British government ventured further into the sponsorship of information technology. It launched the £350 million Alvey programme. Money was to be used for research and development; it was intended to enable Britain to compete with Japan for a new generation of computer systems (Department of Trade and Industry, 1991). Both Inmos and Alvey failed, partly because the British government chose to stand aloof from the projects. By contrast, other countries have evolved a system of direct state intervention in technology policy.

A less direct form of state underwriting occurs in the subsidisation of science. The funding of science should not be seen as a matter of purely academic concern. Scientific research and economic performance are often evoked together. In Britain, it is claimed that the underfunding of science has reduced the country's competitiveness – 'British invention declines as research funds shrink' was one national news headline. But there is a danger of making the connection seem too simple. Norman Clark warns:

It is commonplace nowadays to argue that in some general sense technological changes have had a fundamental impact upon economic growth since the industrial revolution. However, the form this impact has taken and the social relationships involved are even now understood only very imperfectly. In particular, the influence of social expenditures upon science (through R & D, scientific institutions etc.) is problematic (Clark, 1985, p. 3).

Clark does not doubt that *some* scientific advances have improved economic performance, but he challenges the assumption that *all* such improvements are the result of scientific or technical change.

Whatever the actual pay-offs, there is an increasingly close connection between science policy and technology policy. This was not always so, but today the connection is very apparent and it is expressed through the state's role as underwriter. The state comes to play this part, argues Yearly (1988, pp. 109–10), in order to make good the market failures that would result otherwise. The connection between government funding and technological development has a long history. Since the nineteenth century, mining companies have depended heavily upon the state-funded British Geological Survey, which the companies were unable to underwrite but without which the companies could not begin to identify potential sites for excavation (Wilkie, 1991, pp. 11–12). Only the state can provide the resources for research which would be too costly for the individual corporations to bear. How this role is played, though, varies between and within countries. By way of illustration, it is worth concentrating on the single example of biotechnology.

Biotechnology has generated a wide range of political relationships. A 1988 survey of 15 OECD countries found that some governments had developed national plans and programmes for research and development (R&D) work on biotechnology, although others had gone no further than establishing coordinating committees which had very limited powers and enjoyed no opportunity for direct intervention, and some had not even begun to provide public support in any form for biotechnology (OECD, 1988, pp. 15–16). Within these broad categories, there were further variations. While the Germans and the Dutch had developed integrated biotechnology programmes, they had done so at very different times. Germany began work in 1972, ten years before the Netherlands. Countries varied too in the way they focused their biotechnology, some choosing to concentrate exclusively on industrial work, while others set a broader agenda to include medical applications.

Part of the cause of the variations has lain in the political tensions around research funding. The demands of academic research do not fit neatly into the expectations of the state. This issue becomes most acute when military or defence funding is involved. The principles of academic research, which require the free interchange of information, can clash with those of military security, which require secrecy. A similar tension can emerge where commercial competition is involved. We discuss these matters in more detail in Chapter 4. Here we need merely to observe that state sponsorship of science and technology breeds a series of potentially charged political issues, the outcome of which will depend upon how the relationship between the state and technology policy is organised.

POLITICAL STRUCTURES OF CONTROL

To observe the different forms of involvement is to paint only a very general picture of the way in which the state and technology coincide. A more detailed picture involves drawing attention to the political structures which organise the state's role as customer, regulator and underwriter.

State as customer

A key structural difference can be found between those states with highly centralised forms of control and those with a more fragmented system. The contrast is well illustrated by the case of medical technology. Medical technology has been developed faster and deployed further in the US than in Britain. The explanation for this, according to Bryan Jennett (1986, p. 195), lies with the British National Health Service's 'virtual monopoly of health care provision' which is organised according to 'broad central policies' (at least until the late 1980s). Such a structure provided for a degree of planning and a minimum of competition. The centralised structure tended to thwart the duplication and proliferation of medical technology that characterised the US system. Of course, the explanation is not quite this simple. (An equally potent source of control comes from the under-resourcing of the British system.) But within the National Health Service and its central control, there are mechanisms for attempting to assess technology policy, albeit sporadically and in restricted areas (ibid., 195–6). Any move away from central control, to a more privatised, autonomous system is likely to diminish the power over technology policy, and as a result there may be duplication, over-expenditure on technology and inadequate assessment procedures (ibid., pp. 216–17).

Mary Kaldor (1983) draws similar conclusions from her study of the political organisation of defence in the Soviet Union, although this time the effect of centralisation is measured in inflexibility. The political structure appears to have had a direct impact on the form of the weapons system. The Soviet's command structure, she suggests, led to a 'conservative' arms policy. 'The [policy] process', she writes (p. 85), 'discourages technical progress because emphasis on quality or innovative change might disturb the quantitative fulfilment of plan indicators, disrupt established supply lines, and face new supply constraints, and might incur unacceptable risks of failure'.

The Soviet system's resistance to technical change was, in part, a result of the barriers it created to internal and international competition. During the Stalinist period, observes Parrott (1983, p. 74), 'the biases of the mobilizational bureaucracy against slack resources, against professional autonomy and lateral cooperation among specialists, and against the diffusion of technological information, all militated against indigenous innovation'. Bureaucratic stagnation was reinforced by cold war sentiments. Acknowledgement of the quality of American technology, the embodiment of capitalism, was thought to undermine the regime's 'domestic legitimacy' (ibid., pp. 89–90). The result of this fear was to distort not only the industrial structure of the Soviet Union, with its reliance on heavy industry, but also the political structure, through the creation of a dominant military industrial complex (Sakwa, 1989, p. 116). The industrial distortion is bizarrely represented in the fact that, according to Martin Walker (1987, p. 55), 'the best refrigerator that Russians can buy is the Biryuza model, which is produced by the factories of the strategic rocket forces'.

Only when fear of falling behind the West became most acute, in the late 1960s, did the Soviet authorities change the organisation of technology policy. Even then, there was a complex interplay of political forces around the new policy: 'party ideologists and political police had the smallest professional reason to desire more foreign technology. . . . Thus they acted from day to day in ways that impeded the acquisition of large technological benefits.' (Parrott, 1983, p. 293).

Although Gorbachev has initiated many reforms within the Soviet regime, technology policy remains subject to severe structural restraints. As Ronald Amann comments, 'Successful innovations occur but they invariably go against the prevailing tide as a result of political will, individual initiative or, more often, both. It is not by accident that the successful innovator is typically portrayed as a 'hero' in Soviet industrial novels' (Amann, 1986, p. 16). As Sakwa observes, three years into *perestroika*, the problem remains: 'The introduction of new technology is inhibited by institutional conservatism, restrictive price organisation and lack of incentives. . . . The old cumbersome central planning machinery has become a brake on development' (Sakwa, 1989, p. 260). The difficulties seem most acute in the development of computer technology (Glenny, 1988). The corollary is, according to Williams (1990b, p. 215), that innovation 'thrives best in a free and open society'. But such generalisations have to be qualified; there is, after all, no one way to define or organise freedom.

Political barriers to technical development are not exclusive to Eastern Europe or planned economies. Even within market economies, obstacles emerge. One of the key features of Britain's R&D expenditure is the bias towards defence work. Of the OECD nations, Britain comes third (after the USA and France) in the percentage of GDP it commits to defence R&D (see Table 3.1).

Table 3.1   Government funding of defence R&D as a percentage of GDP

Italy	0.08
France	0.52
FRG	0.13
Japan	0.02
USA	0.83
UK	0.43

Source: Annual Review of Government Funded Research & Development 1990.

Britain's commitment to defence R&D owes much to the close links between the Ministry of Defence and arms manufacturers. Arms companies, furthermore, are reluctant to move into civil work because of the relative stability of government demand. One of the arguments advanced by Lucas Aerospace, when it opposed the proposals of its Shop Stewards Combine for an alternative (civil) plan for the company, was that there was a more certain future in government work (Wainwright and Elliott, 1982, pp. 114–15).

Planning in the USSR pushed against innovation because it discouraged initiatives which would disrupt the interdependence of the elements tied into the process. In the West a similar effect is created by the corporatist interests organised around high technology. Government and industry are, as Galbraith (1974) explained, committed to long-term, high-cost work. Under such circumstances, there is an inevitable resistance to change. Change is confined to modifications – hence the 'baroque arsenal' in which a basic design is altered and added to, even where a radical rethink might be more appropriate.

What the Soviet and Western examples suggest is that a political structure which excludes economic competition or political pluralism is liable to have a technology policy dictated by particular interests. The intimacy of the network around weapons technology is a prime instance, and one whose effectiveness was most vividly illustrated in the way the Reagan administration was persuaded into adopting the Strategic Defense Initiative (SDI). This highly costly, uncertain technology came to dominate the political agenda through assiduous lobbying within a closed network. The combined pressure of research-funding organisations like the Hertz Foundation, actors like the nuclear weapons scientist Edward Teller, arms manufacturers and their political supporters, was enough to ensure a decision, the wisdom of which was severely in doubt (Broad, 1985).

Political structures do not divide neatly into open and closed systems. A more open structure may have equally problematic consequences. A lack of central coordination can lead to the underdevelopment of technology policy. Much was made of the Thatcher government's reluctance to fund science and to commit itself to European research projects. Without commitment from the centre, it is argued, there was no support for long-term research projects whose benefits could not be identified in advance. Nor would there be any support for public goods which private sources could not afford. Whatever the value of nuclear power, it is noticeable that the British programme has continued where the US one has largely collapsed. One explanation for this lies in the political structure. The US's federal system does not provide the resources or incentives for a nuclear power programme. Centralism may be conservative in the form of technology that is adopted, but it provides for decisiveness in the face of potential political unpopularity. Such thoughts occurred to the British nuclear industry when opposing the privatisation of the electricity industry. It also occurred to the British Government, which proposed a statutory obligation on all new private electricity suppliers to take a certain percentage of their electricity from nuclear stations.

State as regulator

Political structure does not just affect the development and introduction of technology. It also applies to its regulation and accommodation. Typically, regulation refers to control over the dangers posed by technology, and indeed this is what most of this section concentrates upon. But regulation can also apply to the control of a technology. Where technology is publicly owned, this means establishing appropriate management structures and codes of practice. In the private sector, government control is less direct, and is limited to the establishment of property rights which legitimate control. The most common method for this is through the patent system, by which rights to technical inventions are ascribed, and through which opportunities for the exploitation of them are managed.

Nuclear power technology provides a good example of the ways in which the regulation of technology varies with types of political system and styles of political management. This is most interestingly revealed in much of the discussion of the Chernobyl disaster. On 25 April 1986, the fourth reactor of the nuclear power station at Chernobyl, near Kiev, exploded. Its effects are still being measured in human lives and in its consequences for agriculture and the environment. It is a cost that is being borne not just by Soviet citizens, but by people in neighbouring countries too. There are two features of Chernobyl that are worth observing here. The first concerns the way Chernobyl was run. The second concerns the Soviet government's response to the disaster.

The administration of Chernobyl holds the key to the accident itself. The management structure was inadequate to the task of running the station, but more importantly, the communications network and the ethos surrounding its use were not commensurate with the task of providing an accurate and quick response to the crisis. Douglas Weiner (1990, p. 815) described Chernobyl's staff as working in a politically 'tyrannical and vengeful environment'.

But while the operation of Chernobyl reflected the old political ways, the reaction to the disaster suggested a new era. Some have argued that Chernobyl was both cause and expression of the era of *glasnost*. Initially, the government reacted with, in Martin Walker's words (1987, p. 231), 'the old Kremlin instinct to secrecy'. There followed, though, a change of heart. The Chernobyl disaster was given extensive coverage in the Soviet media and was openly discussed by the political elite. Such a reaction was in stark contrast to the one which had greeted a previous nuclear accident at Khystym. In 1957, a nuclear waste dump polluted a vast area, causing lives to be lost, villages to be evacuated, and food and livestock to be destroyed. But in 1957 nothing was said about Khystym, and it was kept a closely guarded secret for twenty years (Medvedev, 1980). Why did the same type of problem evoke such different reactions?

The answer may seem straightforward. It might reasonably be argued that a highly centralised, bureaucratic structure, through its monopoly on information and its disinclination for self-scrutiny, would not accord safety a high priority and would protect itself by controlling information about any such accident. But notice that in Britain there has been equal caution concerning the release of information about nuclear accidents (Patterson, 1976). 'Centralisation' alone is an inadequate grounding for explanations of government reactions to technical disasters or risks. In their *Risk and Culture*, Douglas and Wildavsky (1982) suggest that different political structures dispose people differently to risk. A closed hierarchy screens out risk, whereas a more open-textured organisation is more aware of the dangers in the world. In other words, the degree to which dangers are perceived is determined by the networks around them. Government regulation is not applied to 'dangers', it is part of the process by which those dangers are themselves constructed. The Bhopal chemical explosion, which resulted when a US company operated a less stringent safety code in India than in America, can be analysed in this way. The 'risks' were defined through the political and economic context which surrounded the technology. Economic dependence may lead to circumstances where stringent safety regulations are not 'seen' to be necessary. This is illustrated in Crenson's comparative study of the air pollution policies of two US cities. One acted swiftly to control pollution; the other was more tardy. Part of the reason for the difference was the *perception* of the problem. Different political formations see problems differently:

> Alterations in these [urban social and political] institutions may produce shifts in the kinds of urban problems that city dwellers choose to complain about. Recent public concern about air pollution may therefore have been generated in part by institutional change (Crenson, 1971, p. 17).

For Crenson, one key variable in the political structure was the business community and its influence.

There is another way in which the regulation process is subject to political pressure. All technologies generate a body of expertise around them. This expertise may itself lead to the creation of powerful vested interests. That, at least, seems to be the case with the development of nuclear power and pharmaceuticals. If the political system allows access to these interests, then perceptions of the dangers and risks may be shaped by those with a strong incentive to minimise certain problems. Writing of the regulation of pharmaceuticals in Britain, Sharp and Holmes conclude:

> In Britain the cosy relationship between ministry regulators and the industry has helped to build an environment in which the industry has grown and flourished, albeit at the cost of relatively high prices. But there is little doubt that it is industry not government that is 'in control' (Sharp and Holmes, 1989, pp. 222–3).

The impact of regulatory arrangements is influenced by the role played by a government in the balance of power. Whereas in Britain the government is often swayed by industrial interests, in France the regulatory arrangements give places to outsiders, which reduces the power of the industry. This is indicative of a general difference between the two countries on matters of regulation. As Erik Millstone (1989, p. 208) observes, 'the French system is more cautious in the face of scientific uncertainties and social conflicts of interest'.

Political structure, though, is only part of the regulatory story. The character of the technology also plays a part, and biotechnology presents a particularly acute set of problems. Not only is the subject richly complex and the available expertise deeply committed to its preservation, but the consequences and problems which it generates are very difficult either to anticipate or to solve. Moreover, even when decisions are reached about the likely outcomes and their desirability, there are further obstacles encountered in giving effect to them. Both the development and the application of biotechnology extend across many policy areas and institutional settings. Biotechnology has relevance for health, agricultural, energy and environmental policy, while its driving forces are to be found among venture capitalists, research scientists, multinational corporations, research-funding bodies, and so on. In short, the political character of regulation policy is a mix both of political structures and the particular features of the technology.

State as underwriter

All states evolve some mechanism for subsidising the development of technology. More accurately, all states create a wide variety of different methods for organising their contribution to technological innovation and implementation. If, for example, we compare their underwriting of scientific research, it is noticeable how countries with roughly similar political sytems allocate different sums. In 1987, Britain spent less (0.58 per cent of GDP) on civil research than Italy (0.70 per cent), West Germany (0.96 per cent) or France (0.91 per cent) (Wilkie, 1991, p. 120). If we compare responses to the same technology, we see a similarly diverse pattern. Take the example of biotechnology. In the USA the key actors are the biotechnology companies; in Japan and France the impetus has come from the state; in Britain it has been left to academics who have campaigned for a biotechnology policy (Sharp and Holmes, 1989, pp. 223; OECD, 1988). Such disparities cannot be explained merely by reference to the political priorities and postures of the respective governments. Instead, we need to focus on the way these priorities are both shaped and expressed through both general strategy and administrative structure.

The British system is very different from, for example, that of Japan. The British system has fought shy of direct intervention (especially in civil technology). There is little attempt at direct control of technology policy, despite the willingness of ministers to talk as if there were such a thing. Although the role of Japan's Ministry for International Trade and Industry (MITI) has been exaggerated by Western commentators, who have seen it as both innovative and interventionist, it has been able to exercise direct control over the development of technology. Richard Ennals has compared the European Community's attempt to organise technological development, ESPRIT, to the system operating in Japan. ESPRIT offers support to industry-led initiatives, MITI intervenes in, and directs, the development of technology (Ennals, 1986, p. 11). Variations in political structure produce differences in technology policy and technology development. Key variables are the degrees of central control and the range of government-funded intermediary institutions. Britain has a closed, narrow structure, but with little centralised direction. Germany, by contrast, has a complex network of institutions but relatively strong central control (Ince, 1986; Wilkie, 1991).

These differences can be ascribed to the policies of the two countries. Margaret Sharp describes the course taken by British biotechnology policy as being like that of a 'rudderless ship, launched but carried with the tide and lacking any positive sense of direction' (Sharp, 1989, p. 156). In Britain, not only has there been conflict between the responsible grant-giving bodies; there has also been no guidance provided by the ministries with a direct interest in the technology, whereas in Germany, biotechnology has been an integral part of a general technology development programme (OECD, 1988).

LIMITS OF POLITICAL CONTROL

So far we have provided a broadly-drawn picture of the ways in which the state has become embroiled with technology. We have observed differences in the general relationship between the state and technology policy. The two main sources of variation are those generated by political structure and strategy, and those generated by the technology itself. In this last section, I want to explore in more detail the problems entailed in managing technology policy.

Just before the 1987 British General Election the government shelved its plans to find sites for low-level nuclear waste in Bedfordshire and Essex. Whatever the official reason for this decision, an important incentive was the unpopularity of the policy. People did not want waste tips near them, even if they agreed that the waste had to be put somewhere. This is another case of the NIMBY (not in my backyard) phenomenon: the refusal of individuals or groups to accept the costs of a particular policy. It does not apply just to technology, but technology focuses the problem sharply. The building of roads or rail links, the dumping of waste and the construction of nuclear power plants induce a sense of fear in local residents. They feel their way of life is threatened by these developments.

Democracy, because it rests on some notion of consent, is faced with a particular problem where such fears are voiced, albeit by a minority. Within a liberal democracy, limits are placed on the power and scope of the state. Not only does this account for the importance of consent, it also means that the state has no clear and incontrovertible concept of the public good by which it can overrule dissenting voices. This creates problems when it has to justify both expenditure on long-term projects and when the costs of those projects fall unevenly upon citizens. The state has to find a means of settling the controversy, and the use of enquiries and commissions presents a mechanism by which to establish the legitimacy of a particular course of action.

But prior to the implementation and legitimation of a technology, the state has to address the question of what policy to pursue. At the level of representative politics, this decision poses an immediate problem. The form of political representation adopted in the West does not typically reward technical expertise. Selection and election is largely determined by party political factors. Thus the representatives who are formally responsible for the policy are unlikely to be acquainted with the intricacies of complex technology. While he was at the British Ministry of Technology in the 1960s, Tony Benn noted in his diary: 'Nuclear Policy Committee, to consider the position on the centrifuge with the talks coming up this week. I had put in tow complicated papers written by Michael Michaels. It took me about four hours to master them, let alone present them' (Benn, 1988, p. 153).

There are, however, ways of easing the tension between representation and expertise. One way is to reduce the part played by representative politics; the other is to subjugate the expertise to political will. The British system tends towards the former strategy. As Ince (1986, p. 187) observes: 'The webs of influence in the higher echelons of British policy-making are so tightly knit that there are virtually no points of entry for outsiders'. Attempts to provide a more open structure have not been especially successful. When the British government was making information technology policy, it was advised internally by the Information Technology Policy Unit and the Chief Scientific Adviser; externally, it had the benefit of the Information Technology Advisory Panel (ITAP), which included industrialists and academics. But despite this broadening of the sources of advice, the outcome was still determined by the internal, executive interests. The civil service used the opportunity afforded by the advisory system to organise a consensus around established Whitehall practice (Keliher, 1990, pp. 68–9). A Treasury official explained their approach to the advice they were offered: 'We wouldn't ask much about the technicalities because we wouldn't be able to judge even if someone told us' (quoted by Keliher, p. 69). Instead, the aim was to obtain agreement rather than understanding. Recording the function of the advisory committees built up around biotechnology, Yoxen (1983, p. 62) writes, 'they were a convenient device for reining in contending parties and condensing disagreement into something useful in a situation of uncertainty'. This tendency to convert advice into terms which suited the organising interests in the policy process inevitably closes off the opportunity for scrutiny and public discussion.

It might be tempting to explain this type of political behaviour as peculiar to the British civil service, with its ingrained suspicion of professional specialisms. But it is worth noting that the problem is not exclusive to Britain, and that it may not even be the worst example. Writing of the US system, Rich (1987, p. 161) observes that 'public officials have been left to rely on intuition and ad hoc procedures'. Holmes and Sharp (1989) argue that France, despite its more centralised state and its cabinet system, lacks both the means and expertise to control technology policy. The French state incorporates departmental rivalries which are not resolved within an effective Cabinet structure and which therefore prevent effective planning. Furthermore, the French recruitment of civil servants from an exclusive educational system works against the appointment of scientifically-trained candidates who have problem-solving or technical skills.

Not all political systems, though, adopt the same postures towards technology assessment as either Britain or France. The most obvious exception is the US Office of Technology Assessment (OTA). Established in 1973, OTA serves Congress by furnishing detailed reports on technologies or technology policies. It employs its own professional staff of 80–90 people, who cover all the major disciplines, and has an annual budget of $18 million. For most projects it creates advisory panels, contracts external researchers and seeks some form of public participation. About 20 reports are issued annually, each taking about two years to produce. The assessment report is an attempt to synthesise the available knowledge on a technology (Wood, 1987). In other words, OTA works to a version of the consensual model, making it superficially similar to the British and French approaches. The difference is that OTA's assessments are better resourced and more influential. Furthermore the consensus is not driven by bureaucratic convenience or norms, but by a desire to agree upon the evidence.

The operation of OTA has, however, to be set in its political context. Its impact is not simply a feature of its institutional form and its resources. Its power is crucially determined by the fact that its client – Congress – has independent authority of a kind which is denied to parliament in unitary states like Britain or France. This also means, of course, that the results of OTA inquiries are themselves subjected to political pressure and interpretation.

But even where the political structures and personnel provide the resources for technology assessment, there remains the question of the criteria. Even with the most superbly qualified individuals and within the most rigorous systems, there is the need to decide what criteria are appropriate. In the case of medical technology, should the assessment be made primarily in terms of cost, or in terms of 'equity' and 'adequacy of access to', and 'efficient use of, restricted facilities'? (Jennett, 1986, p. 1986.) Cost accounting, it is important to stress, is no more straightforward than is determining what is meant by equity and access. Costs have to be measured against savings and results, all of which require judgements. There are, as a result, many forms of assessment, focusing on different aspects of service delivery (ibid., pp. 204–9). In other words the criteria of assessment pose a vast array of problems, whatever political structure is operating. They are not made any simpler by allocating them to the democratic process.

Technology policy tends to work to a longer time horizon than democracy. The relatively short cycle to which democracy is required to work may clash with the longer time scale required for technology. Whilst a particular leadership is in power it may prefer pragmatic, short-term solutions to those long-term solutions which a coherent technology policy might need. It is important, though, not to attribute all 'short-termism' to the democratic process. It also operates within the economy, particularly when immediate profits are valued over investments, when finance capital dominates industrial capital.

Equally, the NIMBY phenomenon may work to distort judgement so that, in a two-party system, decisions taken by the dominant party may owe more to the distribution of votes than rational planning, assuming that the two are not equivalent. Transferring the problem to the voters seems to solve nothing. At one level, most citizens are ill-equipped to understand technology. This is, in part, a legacy of an education system and culture which undervalues scientific knowledge (Wiener, 1982). But the problem is not simply one of ignorance. There is a strong disincentive to acquire the relevant knowledge. This is because an individual knows that their vote or opinion counts for little in any large, complex democracy (Downs, 1957); and also because the urge to acquire information often rests on the existence of a direct interest in the issue (Schumpeter, 1976).

Democracy does not adapt easily to the demands of technology control. The time scale it works to, the expertise it generates, and the interests it incorporates, all conspire against comprehensive control. Robert Dahl remarks on the 'general weakness of the democratic process – failure may be too severe a term – in dealing with highly complex questions, no matter how momentous they may be' (Dahl, 1985, pp. 5–6).

It is worth emphasising, however, that the problems posed by technology for democracy have been set in the context of a particular account of democracy, the liberal variant. A different structure of democracy may not be vulnerable in the same way. It might be supposed that a direct democracy, or one-party democracy, resting upon a notion of the 'general will', will eliminate some of the undesirable effects of party competition. It will also be more able both to incorporate expert advice into the policy process and to allow for a longer time horizon.

These advantages have, however, to be set against the theoretical and practical concerns about direct democracy. We have already identified some of these weaknesses in our discussion of the Soviet Union. The absence of political competition may lead to an underestimation of the risks entailed in any technology. It may also create an inadequate mechanism for both the operation and assessment of that technology (Gustafson, 1979).

Arguments about the general form of democracy cannot be separated from an account of technology itself. The attempt to exercise control has to be set against the problems entailed in controlling any particular technology. The problems of control are greatest if the technology extends beyond established political borders. As Weeramantry remarks (1983, p. 28), 'Like the ancient doctrine of precedent, the ancient doctrine of national sovereignty is inadequate in the scientific age'. The development of information technology encourages such thoughts. Satellite broadcasting, for example, by its nature crosses national borders. It is, therefore, very difficult for a national government to police the content of the broadcasts which its citizens can receive. Insofar as political structures are confined to national boundaries and technologies increasingly cross those boundaries, there are real limits to the possible extent of democratic control. These problems should not be ascribed solely (if at all) to the peculiarities of the technology. Their transnational character owes much to the rise of multi- or transnational corporations who produce the technology and who operate in a global market. Alan Cawson (1989, p. 76) remarks of consumer electronics (VCRs and so forth) that 'the major firms are now beyond the reach of policy measures taken by individual national governments'. This does not make control impossible, but the organisation of transnational controls is notoriously difficult. Even Japan's MITI, the classic case of state intervention, has had its powers reduced by the growth of transnational corporations. And attempts to create a management structure within the European Community (EC) have proved very difficult (Williams, 1989).

CONCLUSION

This chapter began by examining the various ways in which the state has become linked with technology and how it has sought to manage this relationship. We have identified different forms of control (underwriter, regulator, customer) and variations in political structure and technology. We have ended by looking at the problems that both political structures and technological forms pose for the relationship. Together these two general concerns have had a modest ambition: to establish the sort of concerns and issues that must arise in any discussion of the way in which democracy and technology can be combined. There can be no sensible argument unless account is taken of the different types of relationship that may exist or of the different political structures and strategies that may accompany these roles. Only when attention is paid to them can we begin to understand the sort of problems that have to be addressed by any set of political principles that seek in some way to manage technology. Equally important to this discussion is the part played by the changing nature and organisational form of technology itself. We have only referred to this latter aspect in relation to such matters as the complexity and lead-times of technology, and its transnational form. There are, however, other dimensions to technology which have important political implications. These are to be discovered in the creation of technology itself, in the appliance of science.

4 The Politics of Science

Governments, whether democratic or dictatorial, struggle to control technology. Some of their problems, as we have seen, are of their own making; some are a consequence of the activities of other actors. But in concentrating on the formal political mechanisms and the wider economic environment, there is a danger of overlooking a key element in the development and operation of technology. This element, which links the external and internal worlds of technical change, is science. Science policy contributes to the resources upon which technology draws, and the detailed work of science is crucial to shaping technology. We have seen in the previous chapter how political structures and values can play a part in science policy. We now turn to the question of whether the practice of science is subject to similar political processes. Does the politics of technology need to include the politics of science?

In the past, new technologies were often the result of practical experimentation and trial-and-error. It was possible for a development to occur without any accompanying understanding of how it worked, or of the theory behind the technology. The breakthrough in radio and telegraph technology was like this. Messages were being transmitted in the absence of any systematic understanding of radio waves. Such examples are now much rarer. 'Pure science' (as opposed to 'applied science') has been the source of the key technological innovations of the late twentieth century: the bomb, the computer, genetic engineering. All of these entailed the drawing together of discrete pieces of scientific knowledge, each of which may, on its own, have constituted an apparently abstract insight. These insights have subsequently been brought together and exploited by governments and industries for commercial and/or political purposes. The application of abstract knowledge, according to Daniel Bell, forges a distinct new social structure:

> it is undeniable that something substantially *new* about technology has been introduced into economic and social history. It is

the changed relationship between science and technology, and the incorporation of science through the institutionalisation of research into the ongoing structure of the economy (Bell, 1973, p. 196).

Theory now precedes practice; industries are engaged in the application and exploitation of knowledge. Political institutions play a central role in enabling this process to work. But the politics of the relationship are not simply those of coordination; this is not an innocent case of match-making between science and industry, in which the state acts as Cupid. Political values and interests play a part in determining what science is provided for which industry. And not only is politics involved in the particular match made, it is also present *within* science. At least that is the argument of this chapter.

Just because science is an important actor in the development of technology, we cannot, however, leap to the conclusion that it is a key site of special political interest. It can be important without being *politically* important. It might legitimately fall outside the purview of students of politics. We acknowledge, for instance, that all politicians have brains, and that particular neuronal behaviour is causally related to particular forms of public behaviour, but we do not then conclude that political scientists should also study neurology, or that neurosurgeons would make good political analysts. In the same way, acknowledging that science is a key component of technology does not automatically commit us to include science in our concerns. Science only becomes of interest to us if it is shown that it contains, and is shaped by, political interests. Or rather, if science were demonstrated to be politically neutral and independent, then it would be of no interest to political scientists, for the same reason that we are uninterested in, say, the neuronal activities of the prime minister.

Science, then, is important, but our concern is whether it is politically important. It is not enough to show that science changes or develops. What matters is whether science changes according to the political pressures brought to bear upon it or at work within it, and whether the truth it offers is partial and selective in ways that matter politically. What follows, therefore, is a review of different aspects of science, to see what, if any, political dimension is to be found there. Or to put it more starkly, we are testing the claim made by Norman Diamond (1981, p. 32):

In its most basic aspects, the concepts with which scientists organise data and formulate theories, science is inherently political. Scientific concepts are not simply asymptotic approaches to underlying truth. They are products of a particular social structure and may in turn either reinforce or challenge the social status quo.

The rest of this chapter explores such claims by looking at various aspects of science. These are: the scientific method, the concerns of science, the language of science, the interests of scientists, social values and science, and the ideology of science.

THE SCIENTIFIC METHOD

It is sensible to begin this study of the politics of science with the scientific method, the way scientists work, the way scientific knowledge is generated. This is a vast and complex area, and what follows is a much simplified précis (see Richards (1983) for a fuller account). We are concerned only to establish whether it is possible for politics to be present in the daily practice and methodology of science. Does the scientific method incorporate political judgements?

One version of the scientific method has scientists running around collecting facts, out of which emerge theories or natural laws. This is the inductive method. Raw data leads to the formation of theories which explain the relationships between the data. Scientists are neutral observers who respond to what they collect or see. It transpires that the only people who believe this account of science are social scientists, or politicians who use this idea of science as disinterested data collection to legitimate their 'objective' policy proposals. Most scientists would question this account of their work. A competing framework is offered by the hypothetico–deductive model suggested by, among others, Karl Popper (1977). According to this, speculative hypotheses are tested by experiment; the only quality such hypotheses must have is that they stipulate the conditions by which they could be shown to be wrong. Their claims cannot be proved true by this method; they can, however, be falsified. Stephen Jay Gould describes this version of the scientific method as 'debunking', and he explains its operation:

the barrel of theory is always full; sciences work with elaborated contexts for explaining facts from the very outset. Creationist biology was dead wrong about the origin of the species, but . . . creationism was not an emptier or less-developed world view than Darwin's. Science advances primarily by replacement, not by addition (Gould, 1984, pp. 321–2).

A similar view is shared by Peter Medawar who summarises his version of the scientific method:

the purpose of scientific enquiry is not to compile an inventory of factual information, nor to build up a totalitarian world picture of Natural Laws in which every event that is not compulsory is forbidden. We should think of it rather as a logically articulated structure of justifiable beliefs about nature. It begins as a story about a Possible World – a story which we invent and criticise and modify as we go along, so that it ends by being, as nearly as we can make it, a story about real life (Medawar, 1984, pp. 110–11).

Scientists would be reluctant for this sort of description to be interpreted as claiming that science is simply an exercise in imaginative deduction. Max Perutz remarks of his own scientific experience: 'During the first thirty-three years of my own research, imaginative guesswork proved useless: only after my colleagues and I had solved the structure of haemoglobin by X-ray analysis could I begin to guess how the molecule works' (Perutz, 1991, p. 198).

Whatever the emphasis placed on the imagination, this account of science rejects the picture of it as simple fact-gathering. It involves the testing of ideas by experimentation, and the accumulation of knowledge in the process. Good ideas survive as accurate accounts of how things work. Both Gould and Medawar build their account of the scientfic method upon the twin pillars of falsifiability and a real world. Science provides a means by which ever closer approximations to reality can be obtained. Creationism is 'replaced' by Darwinism because it tells a more accurate story about 'real life'. This sort of account of the scientific method is widely shared by practising scientists. It is easy to see why. It grounds their claim to be providing objective (and therefore reliable and valuable) knowledge of the world, while also allowing them a creative role. They are not simply uncovering facts; they are building theories which their colleagues can scrutinise. It may be an appealing account of the scientific method, but it may not be an entirely accurate one.

Critics of this portrait of scientific procedure challenge both the notion of falsifiability and of a 'real life' which underpins it (Feyeraband, 1975). The test of falsifiability, for example, may not exist independently of the theory which is being examined; the relevant evidence will itself be determined by the theory. And equally, the notion of the 'real world' is challenged by those who see 'reality' as the construction of different disciplines and discourses (Foucault, 1980). Science is no different in this respect. It is an exercise in power and political rhetoric, another possible way of ordering the world.

But even if we accept the scientists' own portrait of their work, we see that it introduces the possibility of political influence. If science is a process by which a community of people (scientists) tell stories to each other, then it creeps into the realm of social science. The interactions between members of the society, the kind of stories that are told, the ones that are listened to, all of this will depend on how the participants are organised and what concerns them. Even if there are genuinely objective truths about a real world, there is still the question of which truths are revealed about what part of the world; even if the method itself is neutral, the way it is applied may not be. We may not yet be able to claim that this process is politically significant, but we can agree that the question of science's politics remains.

SCIENTIFIC CONCERN

In his *The Structure of Scientific Revolutions*, Thomas Kuhn (1970) describes the history of science as a series of paradigm shifts. A paradigm defines the working practices of scientists; it represents the shared set of rules and practices which serve to establish that the participants are 'doing science'. Each new era is marked by shifts in what counts as 'science' and what is regarded as the proper concern of science. Kuhn develops the notion of 'normal science' which is bounded by the desire to find solutions to a limited range of problems: 'determination of significant fact, matching facts with theory, articulation of theory' (ibid., p. 34). If this is the typical form of science, then it has quite limited horizons, it aims only at 'puzzle-solving' rather than producing 'major novelties'. For the most part, scientific endeavour is confined to refining the solutions to the puzzles which set the existing paradigm. Working within a paradigm can mean being blind to rival interpretations of the same evidence. In his reconstruction of Cambrian life from fossil remains, Gould explains how an earlier scientist, Charles Walcott, had been influenced by his pre-existing assumptions:

> Walcott 'knew' that *Opabinia* was an arthropod, so the animal had to have appendages on its head. Since he didn't find any, he provided explanations for their absence – either they were so large that they always broke off, or they were so small that they became hidden underneath the head. He never even mentioned the obvious third alternative – that you don't see them because they didn't exist (Gould, 1991, p. 127).

Opabinia in fact belonged to a different category of animal, but Walcott could not see this. His perspective was shaped by a mixture of professional, personal and institutional factors (ibid., Ch. 4). Changes in perception do, however, occur; paradigms are shifted when the recalcitrant evidence becomes overwhelming, resulting – eventually – in the changes in paradigm that describe the history of science.

If scientific endeavour is fixed by an appropriate paradigm, the difference between each paradigm is not so much a matter of truth or falsity, but what 'makes sense'. This can depend on the political sociology of science and on the role of science in society. Science becomes subject to the way 'sense' is defined and by the interests organised around any particular version of conventional wisdom. Einstein's account of the earth being bounded by the speed of light, and the relativity that this implied, stood in contrast to the Newtonian vision of infinity and the absoluteness that it implied. The two accounts were incompatible, but each was to a large extent true in its own right. The triumph of Einstein's theories resulted from further, corroborative, research as well as from the raising of new questions about the physical world. Bertrand Russell wrote of Einstein's work that 'it demanded a revolution in deeply rooted ways of thought' (Russell, 1985, p. 410). But such large pressures for change are often accompanied by the 'micropolitics' of science – the personal, social or institutional interests which are tied to scientific paradigms.

It is important to emphasise that the micropolitics of science are linked to the macropolitics of society. When Galileo argued that the earth circled the sun, his vision did not fit with the dominant view, enshrined by the church, that the earth was at the centre of the universe. Furthermore, although he produced evidence of his claim, by the use of his telescope, it did not constitute proof; it just fitted his hypothesis. Besides, Galileo's method and knowledge were not regarded to be as 'authoritative' as other types of knowledge and other methods (most notably, those of religion). Galileo's account was rejected because the kind of knowledge he possessed was not valued highly (much in the same way that astrology is not regarded as generating valid claims). The social status of science is as important as the way its truths are reached. In Galileo's time, scientific knowledge came well down the scale which was topped by religious knowledge. Galileo's insight only became important when the status of science changed and when the information it generated became socially useful. The acceptance of Galileo's ideas coincided with the need for reliable navigation following the growth of trade between countries.

The validity of these particular interpretations of the sociology of science are not our concern here. Indeed, generalisations about the focus and status of science are notoriously controversial. What, however, is revealed by such analyses is the way in which the social and scientific worlds might interact. The implication of this for the politics of science is that its methods and how it is regarded are not fixed, nor are they established by some objective truth. Rather, the content and form of science are affected by the way science is organised both internally and externally. Looking more closely at science reveals how it is imbued with, and influenced by, political values and interests.

THE LANGUAGE OF SCIENCE

Scientific observations and processes have to be described by those engaged in them, and for this they need a language. In this sense, science is no different from any other form of knowledge – it has to create its own terminology. 'Science is about natural reality', writes Yearly (1988, p. 42), 'but it still depends on human judgement and human conventions'. Furthermore, the terminology has to describe things which are largely abstract. Just as sociologists cannot actually see 'class', so nuclear physicists cannot see atoms. A language has to be invented which describes the processes which their observations or speculations lead them to suppose exist. In this sense, science proceeds by metaphor. What is interesting is why certain metaphors are chosen.

Richard Dawkins provides this arresting opening to a chapter of his book *The Blind Watchmaker*: 'It is raining DNA outside. On the bank of the Oxford canal at the bottom of my garden is a large willow tree, and it is pumping downy seeds into the air'. He continues:

> Those fluffy specks are, literally, spreading instructions for making themselves. . . . It is raining instructions out there; it's raining programs; it's raining tree-growing, fluff-spreading, algorithms. . . . That is not a metaphor, it is the plain truth. It couldn't be any plainer if it were raining floppy discs (Dawkins, 1986, p. 111).

While Dawkins claims that he is giving a 'literal' account of events, it is in fact a (highly plausible) metaphor. Dawkins is drawing the link between the way DNA works and the way computers work. The DNA is the software in the sense that it carries information in the form of a computer programme. Indeed, Dawkins' account of evolution is expressed throughout in the language of information technology. He uses computers as an analogy, and in doing so he is selecting what he counts as the significant aspects of each phenomenon. There are after all an infinite number of similarities and differences between any two items. It is all a matter of what is thought to be relevant to the comparison. The computer comparison may be a good one, but only because we find it plausible and enlightening; not because it is 'literally' true. Political theorists have sought to ground their theories in nature for a similar reason. If we think of nature as wild and threatening, then the idea of the 'state of nature' fills us with dread, and a political system that enables us to avoid it will appeal to us. This argument by metaphor depends on the way we think about nature. So Dawkins' argument rests on what we think about, and how we understand, information technology.

Hilary and Steven Rose (1970) give a further twist to the use of metaphor in science. They describe the way in which, for example, the internal workings of the cell have borrowed different analogies in order to make sense of the process. This, though, is not the end of the story. The choice of analogy is not arbitrary or neutral, but rests upon a set of assumptions about ordered behaviour in general. The Roses observe how the language of science changes with time: the way a Keynesian economics description of the behaviour of the cell was replaced by a cybernetics account. These shifting metaphors reflect a complex connection between science and society. On the one hand, scientists draw on the conventional wisdom of society to describe scientific processes, but then those scientific processes come to be used to legitimate social arrangements. The most obvious examples of this are to be found in natural history. Animal behaviour is often described in the language of human society: animals court each other; they are aggressive, they stake out territory. Not only do these analogies draw on human society; they reproduce certain assumptions about that society: about how men and women relate to each other. What can happen is that the accounts of the animal world which are drawn from the human world are returned back to that world and used to legitimate the very relationships that provided the original metaphor. It becomes 'natural' for men and women to behave in certain ways. But there is no compelling reason for believing that animal and human behaviour is directly analogous (Midgley, 1980). The language of science, in drawing on the social world, is not politically innocent. Science not only reflects our culture, it also helps to create that culture.

SOCIAL VALUES AND THE SUBJECT OF SCIENCE

The impact of culture on science is even more apparent in the tasks which science sets itself, or is set. If, as Kuhn (1970) argues, the history of science cannot be seen as a straightforward story about the progressive acquisition of knowledge about the world, then it is reasonable to suppose that the questions which face science are selected. They do not just sit there on some eternal exam paper. One way of illustrating this is by considering the way social values enter into the daily practice of science.

Animal liberationists, who attack the laboratories where experiments are carried out on animals, challenge the prevalent assumption that animal life is worth less than human life. Why is it legitimate to kill animals in order to find (or test) cures for diseases? What makes this more or less acceptable than the use of animals to test cosmetics? And how much does the choice of a particular animal for use in experiments derive from our sense of the animal's cultural importance and how much from its physiological characteristics? Experiments on mice are all right; but on dogs and cats they are not. In short, a set of judgements, themselves derived from a set of political, moral and cultural values, are applied to the way science is done.

Social values also shape the questions which science tackles. It is not, after all, immediately obvious why we should want to correlate intelligence and race, and yet this very concern preoccupied a number of scientists in the nineteenth and twentieth centuries (Jensen, 1969; Eysenck, 1971). Equally, it is noticeable that much more research goes into cancer than into other diseases. Part of the reason for this is that cancer is the current concern of particular populations. The priority given to cancer has a socio-political explanation, rather than a medical one. Social and political assumptions infuse science, and identify the problems that 'have' to be solved.

The interaction between science and society is a factor in this process. The distinction between pure and applied science is becoming less and less clear. This is a function of the way science funding is linked to the state and the industrial structure it serves. Where science may once have been pursued disinterestedly, it has become increasingly difficult not to identify some clear social purpose in the work being carried out (Greenberg, 1967). 'Knowledge for its own sake' is a hard case to sustain as a basis for justifying state subsidy. Social usefulness has to feature in the justification of support. This inevitably creates a pressure on science to produce work which can be claimed to serve a recognised social function. It also helps to establish the case for the democratic control of science.

SCIENTISTS AND THEIR INTERESTS

Sometimes the interests and practices of scientists are directly
determined by the operation of political priorities. This happened
both in Stalin's Russia and in Nazi Germany. Under Stalin, genetics
was a deeply sensitive subject. The very idea of inheritance
threatened the dogmas of historical materialism which held the
environment to be determinant. As a result, work on genetics was
heavily circumscribed; Western advances were not published and
only certain Soviet work was acceptable (Hales, 1982, pp. 26–9).
Under the Nazis, science and scientists were subject to an equivalent
form of political control. Goudsmit (1985) describes how the SS
vetted scientists and ordered their allocation. Scientific endeavour
was directed to serving Hitler's will. This involved the creation of a
super-gun, which, it turned out, could not be made to work. No
one, though, had the courage to inform Hitler of this, with the result
that work on the construction of the gun continued. More
horrifically, scientists became involved in inhumanly cruel experiments in the concentration camps, subjecting captives to all manner
of 'tests'.

These extreme examples have to be set against other ways in
which science and politics have become enmeshed. It is not just the
assumptions of societies or politicians that affect the priorities with
which science operates. The interests of scientists are also tied up in
the character of science. Gould recounts the case of Samuel
Morton, a nineteenth century scientist who tried to argue that
races differed in their intellectual capacity. His measure for
intelligence was the size of the brain, and his method for obtaining
data on this was through examination of his vast collection of
skulls. Gould explained Morton's procedure:

He [Morton] filled the cranial cavity with sifted white mustard
seed, poured the seed back into a graduated cylinder and read the
skull's volume in cubic inches. Later on, he became dissatisfied
with mustard seed because he could not obtain consistent
results.... Consequently, he switched to one-eighth-inch-diame-
ter lead shot 'of the size called BB' and achieved consistent results
that never varied by more than a single cubic inch for the same
skull (Gould, 1984, p. 53).

Morton had published all his data, and Gould found that 'Morton's summaries are a patchwork of fudging and finagling in the clear interest of controlling a priori convictions' (ibid., p. 54). Unlike Morton, Gould found '*no* significant differences among races for his data'. Importantly, Gould also reports that he found 'no sign of fraud or conscious manipulation. . . . All I can discern is an a priori conviction about racial ranking so powerful that it directed his rankings along preestablished lines' (ibid., pp. 67–9). Political values explain Morton's results better than the method used.

This is but one way of illustrating how science, scientists and their interests may become entwined. Their interests may take a variety of other forms, beginning for example at the level of career prospects. Promotion depends first on results, and then on generating results in areas which are designated as important. Meeting these standards puts considerable pressure on scientists, who have on rare, but highly publicised, occasions actually faked their results. Others, in a desperate bid to be first, have made claims that later could not be validated. In the late 1980s, two scientists claimed to have created nuclear fusion at low temperatures and called a press conference before publishing their results. Although many have replicated their experiment, nuclear fusion did not occur, so the claims of the two scientists have been treated with great scepticism. More typically, the scientific agenda is set by the distribution of power within the profession. The work that is done may not constitute that which is most useful or interesting, but rather the work that promises funds and promotions.

The internal politics of the profession can give way to a wider set of interests. Andre Gorz (1978) has argued, adding a Marxist gloss to the arguments of Daniel Bell, that scientists have a distinct class interest and that science is used in pursuit of it. To illustrate the ways in which science can become shaped by social interests, I shall take two case studies: women in science, and biotechnology.

Case study 1: women in science

Brian Easlea, in *Fathering the Unthinkable* (1983), suggests that the gender of scientists is an important factor in accounting for the work they do and the character of science itself. For Easlea, the creation of nuclear weapons is a direct consequence of a male dominated science. There are a number of stages to this argument. The first concerns the employment predicament of women in science; the second considers science's concerns and their relevance to women; and finally there is the issue of whether science as a discipline favours men over women.

Women as employees in science

For women, the opportunities for a career in science are much less than for men. In the US and Britain women are grossly under-represented in science and there is evidence that their chances of reaching the top are very slight. The explanation for this often lies in the exercise of discrimination by men, and by the social pressures which make it difficult for women to pursue a career on equal terms with men. Irvine and Martin (1986, p. 98) conclude their study of women in radio astronomy: 'differing social pressures on men and women scientists must be considered responsible for preventing women from achieving their full career potential. Early promise gives way to the frustration of unfulfilling jobs and any initial brilliance fades quickly into obscurity much in the manner of a shooting star'. Without wishing to downplay this argument, it is important to note that the status of women in science in the US and Britain is not replicated everywhere (Johnson, 1986, p. 105). The point remains, however, that there is substantial evidence of discrimination against women in science, and that it is a discrimination which is even more pervasive than in other professions.

Women and the concerns of science

The second stage in the argument focuses on the question of whether the absence of women in science is a consequence of the focus or the concerns of the discipline. Has science been 'hijacked' by men to serve their own interests? As Walsh observes (1980), biochemical birth control was at first slow to develop and then took a form which failed to recognise the interests of the user. The contraceptive pill is in some ways a peculiarly inept form of birth control. A localised process – conception – is prevented by a technology that affects the entire body and which has a number of potentially hazardous side-effects. The fact that the Pill takes this form, it is argued, is because of the interests which dominate the organisation of science. If men could become pregnant, the argument runs, then a reliable contraceptive device would have been discovered earlier and it would not have been of a type which incurs considerable risk to the user. A similar argument is deployed to explain science's considerable involvement in military research. The general thrust of these claims is that science is not autonomous and that the work it does (the problems it solves) are determined by dominant (male) interests.

Science as a male discipline

A third aspect of the argument about how men dominate science focuses on the discipline itself, on the routine practice (and not just the products) of science. Rose and Rose (1982) write: 'Science is authoritarian, masculine and arrogant in the certainty of the understanding of the world it produces'. Easlea echoes this sentiment when he argues that science, and physics in particular, has acquired a 'masculine' image. He claims that

> physics is widely perceived to be strongly masculine not so much because it is one of the most male-dominated of the sciences but rather because it is widely believed that physics can be successfully pursued only by *male* human beings and, moreover, only by men possessing strongly masculine qualities (Easlea, 1986, p. 134).

In identifying these 'masculine qualities' Easlea is not claiming that they are 'facts' about women and men, but that they are strongly reinforced cultural definitions of the roles which women and men are expected to play. The kind of attributes he cites for women are: 'softness, connectedness, receptivity, the capacity to nurture'. By contrast, men are supposed to display 'aggressiveness, hardness, self-control, disconnectedness' (ibid.). Easlea then goes on to claim that there are six ways in which physics comes to embody 'masculine' characteristics, thereby disposing it to represent a particular set of material and intellectual interests.

1. Members of the discipline are expected to engage in aggressively competitive behaviour; this is the route both to knowledge and promotion. In such a world women do less well. But as Easlea observes, such behaviour is not exclusive to physics, and many other disciplines share this feature.

2. Physics is distinctively masculine in its remoteness from living entities, unlike, say, biology. This distance is combined with an emphasis on control and manipulation, and a strong motivation to establish scientific laws. Hacker (1983) describes this process at work in the 'mathematization of engineering'. The result is a discipline less hospitable to women.

3. The image of the scientific method as an objective and unemotional examination of logic and facts is particularly prominent in physics. Insofar as 'emotional reticence' is deemed a masculine trait, the methods of physics are seen to be more adapted to the cultural norm of men than women.

4. The object of physical research is nature, the identification and analysis of 'laws of nature'. But, Easlea observes, the link between science and nature is loaded with much cultural baggage; in particular, both 'science' and 'nature' are gendered. Science is, according to the preceding three points, male in character; nature, on the other hand, is typically viewed as female. This is indicated by reference to 'mother nature' and so on. The relationship between science and nature is then laden with sexual imagery, in which science uncovers the mysteries of nature or tames nature. Nature is both feared and hated, much in the way misogynist culture fears and hates women. Science as the controller of nature becomes imbued with the aggressive sexual imagery that often characterises relations between men and women.

5. Following from the argument that sexual imagery and attitudes are buried in the portrayal of the link between science and nature, the goal of physics, to conquer nature, also takes on a similar character. Physics becomes the expression of the desire to dominate nature and subjugate it to man's will.

6. The full realisation of this male-driven desire to conquer female nature is to be seen in the development and creation of nuclear weapons, the ultimate expression of man's conquest over nature (Easlea, 1986, pp. 134–50; also Pacey, 1983, pp. 104–5).

There are two questions which prompted the masculine character of Easlea's portrait of physics. The first is: 'why do so few women study physics?' The second asks: 'how should we explain the way physics is organised and the work it does?'. The answer to both questions lies in the cultural characteristics of the discipline. If physics was not so aggressive, manipulative, so concerned to dominate nature, it would represent a more attractive career choice. And, by implication, were women to be properly represented in science, there would be a marked change both in the projects which physics sets itself and in the way they are approached.

How we view this general argument depends upon how we view a number of its features. There is clearly a need to establish the validity of the portrait of physics. Is it aggressively competitive? Is its methodology based upon a suppression of emotion? Does it seek to dominate or manipulate nature? Are the assumptions made about the distribution of masculine/feminine characteristics accurate? Does having these characteristics dispose a person to act in one way or another, to reject physics or embrace it? There are clearly grounds for argument in each case, arguments which are fuelled by the fact that 'physics' cannot be easily compartmentalised. The emergence of alternative schools of physics (Davies, 1984), and the development of the chaos theory (Gleick, 1988), suggest that physics takes a variety of forms and has no one fixed perspective on nature. Equally problematic is how the arguments about gender might be given content: how might we identify 'connectedness' and 'disconnectedness', or emotional or unemotional responses?

But even if we qualify Easlea's particular arguments, we need to remain sensitive to his general concerns: to show that a scientific discipline does not exist autonomously and is subject to many 'non-scientific' interests. And, therefore, these features of science have important implications for the development of technology (Wajcman, 1991, pp. 1–25). The second case study adds to the plausibility of this general hypothesis.

Case study 2: biotechnology

Biotechnology has pushed the issues we are concerned with here into greater prominence. In some ways, biotechnology is less a combination of science and industry and more a political order. Martin Kenney (1986) talks of the emergence of 'The University–Industrial Complex', directly recalling the military–industrial complex of the post-Second World War era. Industry and academia, it is contended, have been pushed into a new relationship in which the interests of scientists and the pursuit of science have been incorporated directly into the process of generating commercial success.

At the heart of biotechnology lies the discovery of recombinant DNA (rDNA). It is this molecule which enables new products to be generated through the insertion of modified DNA into a host cell. The discovery of rDNA in the early 1970s was not motivated by any particular view of its commercial or practical potential. The lack of commercial interest, suggests Kenney, was a legacy of the break between biochemistry and molecular biology that emerged in the 1930s. Biochemistry had been 'captured' by the medical profession in the early twentieth century, and from then on had been an applied science. By contrast, molecular biology was more theoretical in orientation and uninterested in the practical application of knowledge. Hence it was that the structure of DNA was unravelled, not by biochemists, but by two theoreticians, Francis Crick and James Watson.

When the commercial possibilities of rDNA were recognised by molecular biologists, there was, therefore, no particular reason why its application should follow the commercial route taken by biochemistry, that is, through medicine and the major pharmaceutical corporations. Instead biotechnology took the path traced by the computer industry. Like the small-scale entrepreneurs who forged the main breakthrough in computer hard- and software, the biotechnology industry was founded upon modestly sized venture capital enterprises whose products were then taken up by larger corporations. There was, though, an important difference between the embryonic computer and biotechnology industries. Whereas the leading figures in the computer revolution were bright graduates or disillusioned company workers, the key actors in the biotechnology business were established academics. The new industry was built around the shared expertise and interests of universities and industries. University researchers found themselves either acting as consultants for the new industry or running it as proprietors of the emergent companies. There is the example of a Professor at Harvard who owned 208 000 shares in a biotechnology company which also paid him a grant of $333 722 for his research. In exchange the company acquired 'a nonexclusive, worldwide license to commercialise any patents from the work' (Kenney, 1986, p. 151). Academics have also become part of more formal links between universities and industry; they are now being sponsored by their industrial partners. Here lies a key aspect of the politics of biotechnology.

'In the emerging field of biotechnology', writes Kenney (ibid., p. 29), 'the lack of corporate expertise led to unique new arrangements between industry and corporations at institutional levels that are affecting a number of the universities' traditional values and norms'. At one level there was the political question of who had rights to the knowledge upon which biotechnology was based – the universities, or the individual researchers? Equally, if corporations were to buy academic expertise through the sponsorship of departments, who was to make appointments and on what basis? Corporate involvement, according to Kenney, threatened to change the basis of scientific endeavour:

> The earlier fragile system of peer pressure and 'old boy' networks is collapsing under the assault of commercialisation. Behaviour that is normal in industry or the professions – secrecy, evasiveness, and invidious competition based on pecuniary motives – threatens to disrupt the social relationships based on noncommercial motives that are expected of and have characterized the university (ibid., p. 131).

Professors have become tied to companies and their right to move has been restricted because they now deal in trade secrets (ibid., pp. 153–4).

A similar political issue has emerged over academic research sponsored by military interests. In Britain, a political row broke out at Bristol University's Veterinary School. Work that had previously been financed by the Ministry of Agriculture was moved to the jurisdiction of the Ministry of Defence, and some of the scientists involved were concerned that their work was now part of a research programme connected with biological warfare. They protested against the switch of sponsor, but the University authorities argued that no political issue was involved: 'Defence research – any research – can be used to attack, harm or kill people. The work is neutral; the decision to use it in a particular way is a political one not taken in the University, sometimes not even in this country' (*The Independent*, 9 August 1990). The distinction between the 'work' and its application is, if the example of biotechnology is any guide, becoming increasingly difficult to maintain. Neither science nor universities can claim to be above politics nor to be independent of the uses to which research is put. They are both integral parts of the technology system.

SCIENCE AS IDEOLOGY

The politics of science cannot be confined to its practices and its concerns. It must also include the way in which the idea of science itself is used and perceived. Science has an ideological as well as a political dimension, as was demonstrated by the link between the language of science and the practices of society. Science is not just a set of ideas, it is also a set of activities. The idea of 'science' carries with it a number of symbols and values, all of which serve to legitimate both science and a whole range of other activities. 'To be scientific' is often taken to mean to be objective. A scientific approach is held to be more rational (and by implication more accurate) than an emotional or value-laden one. The image of the white-coated expert on TV advertisements (particularly for brands of washing powder or household bleach) is often taken to represent, however implausibly, the voice of disinterested expertise.

Science acquires this meaning in a variety of ways, but there is no doubting the part played by the media. In her study of press coverage of science and technology, Dorothy Nelkin observes how scientists are typically portrayed by the press as 'divorced from normal activity'. It is an image that is further enhanced by the treatment accorded to winners of Nobel prizes. They are made into stars, on a par with the idols of the pop world (Nelkin, 1987, pp. 15–21). Alternatively, other media, most obviously the cinema, have propagated an equally distorted myth, that of the mad scientist – from Dr Frankenstein to Dr Strangelove, from *Back to the Future* to *The Fly*. Whichever image of science is projected, the media is intimately involved in creating the accompanying myth.

Popular presentations also tend to give a distorted impression of the practice of science, presenting it as an Olympic race for a solution to a recognised set of problems (ibid., pp. 6–7, pp. 51–2). And the reporting of the competition tends to further obscure the underlying process: 'Applying naïve standards of objectivity, reporters deal with scientific disagreement by simply balancing opposing views, an approach that does little to enhance public understanding of the role of science' (ibid., p. 68). But the key issue is less the misreporting of science (although this is important); instead it is the general impact of press coverage on the image and ideology of science. 'The reporting of technology, like that of science', writes Nelkin (p. 173), 'tends to be promotional. Many writers convey a fervent conviction that new technology will create a better world'. This general impression is achieved through neglect of the details of the scientific method, its hit-and-miss character, the issues of responsibility and control that it raises. The net effect is to create an image of science which has, on the one hand, a powerful appeal (it can solve our problems), but which, on the other hand, obscures the actual interests at work in its practice.

The power of the feminist critique of science derives, in part, from the prior assumption that science occupies a core role in society. The repercussion of science's biases are felt widely. Understanding the politics of science is important because of its integral place in contemporary society, where it plays a vital ideological role. The feminist critic sees science as formulating a view of the world which marginalises women and alternative knowledge claims. The feminist argument is part of a more general critique of science. According to this, science is used to screen out certain forms of life (Fay, 1975, Ch. 2; Marcuse, 1968). Science constitutes a particular ideological response to the question of how we should live. Science provides a basis for controlling human behaviour, via a crude reductionism. In *One-Dimensional Man*, Marcuse (1968) argued that the scientific method was used, first, to dominate nature, and then to dominate humankind. Science and technology provide the ideological legitimation of 'expanding political power'. Behaviourism, sociobiology, management science and political science are all seen as the children of science-as-ideology (Adorno and Horkheimer, 1979; Fay, 1975; Macintyre, 1981). It is a theme which continues in arguments about post-modernism (Foucault, 1980; Baudrillard, 1988).

Such views of the ideology of science are not universal, however. In his play *Life of Galileo*, Brecht celebrated science as the foundation of democracy. Science gave grounds for doubt, and with doubt emerged the right to question authority.

> Creating knowledge for all about all, it aims to turn all of us into doubters. . . . Our new art of doubting delighted the mass audience. They tore the telescope out of our hands and trained

it on their tormentors, the princes, landlords and priests (Brecht, 1980, pp. 107–8).

Brecht expressed faith in an ideal, uncolonised science which could arm us to expose cant and false gods. For Brecht, science was liberating, but others see it as providing a cloak for tyranny, for the contraction of choice rather than its expansion. But both sides of the argument agree, however, that science plays an ideological role.

CONCLUSION

The various aspects of science we have reviewed above all testify to one basic theme. They all indicate that the character and form of science is not fixed; more importantly, that it is riven with political values and interests. Three scientists summarise their understanding of the link between science and society:

> at any historical moment, what passes as acceptable scientific explanations have both social determinants and social functions. The progress of science is the product of a continuous tension between the internal logic of a method of acquiring knowledge that professes correspondence with and truth about the real material world, and the external logic of these social determinants and social functions (Rose, Kamin and Lewontin, 1984, p. 33).

Aspects of this tension have been the focus of this chapter.

What are the implications for the control of technology? Firstly, it is clear that the extent of the control must include science; secondly, that the form of control must recognise a number of elements in what constitutes science: from the scientists themselves to the ideological rationale behind science; and thirdly, that the organisation and personnel of science are only part of what determines how science works and what it studies.

If we take these claims seriously, then control of technology depends on, among other things, the degree to which science can be subject to control. There is, of course, no simple connection between scientific change and technical change. Scientific innovation does not lead automatically to new technologies, if only because the same breakthrough may invite any number of different applications. It is only with hindsight that we can see how scientific insights contributed and combined to produce, say, the atomic bomb.

Also limiting the control of science is the nature of science, which works against any idea of total control. Although political influences do bear upon science, we cannot take this to imply that if we control the political processes we can control scientific development. The evidence of attempts to impose political control over science – in Stalin's Russia and in Nazi Germany – removed the conditions necessary for proper scientific inquiry. The hypothetico–inductive method of science requires some degree of autonomy. Discovery cannot be dictated. We have to be cautious, though, about the conclusions to be drawn from this. Don Price, for example, sees it as evidence of the uselessness of democratic control:

> The notion of democracy, or ultimate rule by votes of the people, is simply irrelevant to science. For science is mainly concerned with the discovery of truths that are not affected by what the scientist thinks or hopes; its issues cannot be decided by votes (Price, 1965, p. 172).

This argument is weak. First, it starts from a view of scientific inquiry which, as we saw earlier, would not be accepted by scientists themselves. What scientists think does matter. Secondly, even if science poses complex problems which can only be resolved by those who understand the difficulties, it does not follow that democratic processes are redundant. Democracy can operate in conjunction with expertise, and in matters concerning the direction and character of science there seems to be a perfectly legitimate role for voters.

The evidence of this chapter suggests that not only is there direct political intervention in the funding, organisation and priorities of science, but that also the scientific method is itself linked to political values and judgements. Insofar as science is part of the development of technology, then a politics of technology must entail a politics of science. It is time now to look more closely at the substance of this politics, to see how technology has an impact on political resources and interests, to see what might inspire the desire to bring it under control.

5 The Political Effects of Technology

INTRODUCTION

Even the most basic technology can have devastating effects. As the magazine *Small World* (1990) reported: 'Most of the women in Guatemala have an operation to remove their tear ducts when they are fourteen or fifteen years old because smoke levels in their kitchens destroy their natural ducts'. It is such consequences of technology that are the concern of this chapter.

Understanding the way in which technology is developed and implemented, important though this is, tells only half the story. We would be little interested in technology were it not for its impact upon the way we live. The political argument about technology is largely concerned with the way its effects are perceived and judged. It is not just the technological determinists who are allowed to believe that technology shapes the relations between people and the opportunities they enjoy. There are, after all, virtually no activities that are possible without the aid of some technology. It is, therefore, important to consider the kind of impact which technology can have in shaping those activities, to see the ways in which it can affect the conditions and character of the political choices and practices involved.

Langdon Winner (1985) once asked 'Do artefacts have politics?' His answer was 'yes'; he showed how political values and interests were promoted through the effects of technology. He argued that technology can be used to reinforce or implement a form of discrimination. For example, a New York bridge was built too low to allow the passage of buses, so that only those with cars (the rich) could gain entrance to the park that lay beyond. The architecture of most houses and public buildings works in a similar way. Their design and their features make them suitable for only particular uses or people. They are, for instance, not planned to be energy efficient; their inhabitants are forced to heat them with energy from non-renewable resources. Nor do such buildings typically provide access to those with mobility difficulties. Few buildings are designed to accommodate wheelchairs. In these instances, technology becomes the instrument of particular political forces or interests. Design makes life more easy for some people than others; it provides pressures for some practices over others.

Winner goes on to identify a more subtle method by which technology shapes political practice. Technology can cause the existence of a specific set of political arrangements. Winner uses the case of nuclear power, arguing that, because of its potential dangers, it requires the introduction of special rules, regulations and organisation, which in turn need restrictions to be imposed upon society. The potential risks entailed in the transporting and use of radioactive materials *require* the introduction of vetting procedures for employees, restrictions on the availability of information, and the deployment of special (often armed) police forces to guard the materials. The special features of nuclear energy technology are also used to justify the removal of normal employment rights, in particular the right to strike.

In asking whether artefacts have politics, Winner suggests that the introduction of a technology shapes the surrounding political structure and serves (or subverts) certain political interests. Technology, by this account, has a direct political effect. Winner writes as if technology is an independent political actor, and in doing so loses sight of the processes that bring that technology into being and regulate its activities. Nonetheless, his argument forces us to consider the ways technology can aid and abet political interests. My aim here, therefore, is not to challenge Winner's general approach, but rather to tease out its implications for understanding the effect of technology on political processes.

The effects of technology can be measured in a number of ways. They can be identified in people's *dependence* upon technology, a dependence that is brought home every time there is power failure. The loss of electricity, especially on a winter's evening, immediately demonstrates our dependence on it, its centrality to 'normal life': no television to watch, no microwave or electric oven to heat food, no electric kettle, no light to read by, no electric water heater for a bath or shower, and so on.

Such dependence is closely allied to *inequality*, since dependence rarely allows for equal shares. Those who control the technology upon which others depend, exercise power over the users. Similarly, those without access to the technology, those unable even to depend upon it, are excluded from doing those things which the technology allows. Television, it is sometimes suggested, can be understood in these terms. People depend upon television for their news and entertainment, but they have very little control over its content and form. Those without television are in some important respect disenfranchised from the culture which is disseminated through its channels.

The effect of technology can also be seen in the way in which technology can both increase and limit the *choices* available to individuals or groups. Transport technology increases the number of places which tourists can visit; the introduction of factory automation reduces the initiative available to individual workers. Equally, technology can seem to affect political processes through the degree of *control* or *participation* it allows to individuals or groups. Data collection technology, for example, can on the one hand increase the potential for monitoring the activity of citizens by expanding the information held on people, while on the other hand it may increase the capacity of those same citizens to scrutinise the activities of their political leaders through the use of electronic polling.

Technology can alter peoples' *life chances*. It can increase the risks they face by introducing new dangers into their lives – from nuclear weapons to fast cars. Alternatively, technology can improve their prospects of survival, by using medical technology, for instance, to enable them to live longer.

Our *perception* and *experience* of the world and its inhabitants is changed by technology. Since the advent of the cinema we have perceived things differently. Technology not only increases our access to knowledge and culture, it can actually alter the character of both. The development of sound recording technology has allowed us to hear music in new ways and to acquire different aesthetic criteria. With the development of sampling technology, music can be made without recourse to traditional instruments or individual players or composers.

The effects we have identified above have one thing in common; they are a direct consequence of the way the technology has been applied. The effect can be anticipated from the function of the technology, even if this is not always the function which the manufacturer might claim for it. Although washing machines are often advertised as decreasing domestic labour, this may obscure the fact that previously such work was done elsewhere by other people or that the time not taken up by washing is occupied by other domestic tasks (Wajcman, 1991, pp. 93–5).

There are some effects which cannot be divined directly from the technology's function. These are its *side effects*, which are a vitally important aspect of any understanding of the operation of a technology. The most obvious side effect is that of pollution, which threatens life chances, distributes burdens unequally and so on.

These are, then, some of the ways in which technology can be said to have an effect. They do not constitute self-contained categories, and some may overlap (instances of inequality are often expressed in terms of lack of participation). Furthermore, the same technology may easily have a number of different effects. It helps, nonetheless, to keep these effects separate in order to explore them in more detail and to draw out their implications for the politics of technology.

DEPENDENCE

One of the most obvious political effects of the spread of certain technologies is the dependence relationships they create. Technological dependence can be seen both within countries and between them. Dependence can apply to a nation, group or individual. Where it occurs, the dependent actors need a technology in order to survive; they do not, however, have control of the provision or servicing of that technology.

When a technology is taken from one context and placed in a very different one, its operation and its effects can be completely transformed. Dunn gives an example:

a small industry in a North African country . . . was engaged in producing leather sandals for local use. It was suggested that a cheaper product could be obtained by changing from the traditional material, leather, to plastic. The country imported two plastic-moulding injection machines at a cost of $100,000. This highly automated plant operated very successfully, but the net effect was to put 5,000 leather shoemakers out of work and in turn to reduce the incomes of the makers of leather, glue, thread,

fabric linings, tacks, dressings, polishes, hand tools, wooden lasts
and boxes, all of whose livelihoods were connected with this
industry. In their place were just 40 injection moulding operatives
(Dunn, 1978, pp. 16–17).

Indeed, the full effect of this simple change in technology was a
general worsening of the local economy and an increased dependence on the countries that supplied the machines and materials.

Dependence creates an imbalance of power in which the supplier
of the technology can manipulate the recipient. Such a relationship
lies behind the different safety standards that apply to the same
technology in different countries. This has often been the case with
medical techniques which are banned in the West, but are used
outside the industrialised world. In the aftermath of the Bhopal
disaster in India in 1984, when a pesticides plant exploded and
killed 3323 people, made 26 000 chronically ill and blinded and
injured over 200 000 more, it was revealed that the American owners
imposed far more rigorous safety standards in their domestic plants
than operated elsewhere (Weir, 1987).

The problem lies, of course, not so much with the technology
itself as with who owns and controls it – where the profits or
benefits go. The attempt by nations to gain control of foreign owned
technology is an attempt to alter the criteria by which it is operated
and the distribution of the benefits from it. Nasser's seizure of the
Suez Canal in 1956 is a famous example of an attempt to change the
balance of power around a particular technology.

The corollary of the dependence created by technology transfer is
the vulnerability which results from the *denial* of technology.
Typically this applies to military and security technology – for
example, the West's attempt to prevent Iraq from building a 'super
gun' and developing a chemical weapons arsenal by imposing an
embargo on the shipment of certain equipment and materials. But
denial can extend to other technologies, such as computers. In the
early 1980s, the US Congress tried to prevent the movement of
computers into Eastern bloc countries by imposing tight export
restrictions on American technology, even where it was owned by
other nationals.

The impact of denial can be to produce a double subservience.
The first stems from the relationship between the supplier and its
would-be customer, in which the latter has to rely entirely on the

former. The second lies in the predicament caused by the absence of the technology. There may be good grounds for denying technology to states, but this does not change the fact that the denial has a direct and identifiable effect. The repercussions from the distribution of technology, whether in the form of transfer or refusal to supply, can be felt in the balance of power between and within countries. There are, of course, a complexity of ethical and political issues raised by this general process, and it is not clear how and when the transfer of technology should be permitted (Iannone, 1987, pp. 273–304). We do not need to review that debate here; we need merely to observe that measuring the effects of a technology has to take account of the consequences of both having and not having it, and to recognise that the 'effects' are not an objective feature of the technology, but are derived from the context in which the technology is (to be) used.

INEQUALITY

The Bhopal explosion in India revealed how the same event may not be experienced in the same way by all whom it touches. Women suffered a great deal as a result of the disaster: 'the effects on the lungs were made especially severe for women because of the nature of their household responsibilities. Cooking in particular became an even more damaging, unhealthy and unbearable activity' (Simms, 1990, pp. 26–7). In short, the disaster exacerbated existing inequalities and was, therefore, differentially experienced. Substantial inequalities are likely to result from the commercialisation of space. The development of satellite technology, for military and civil purposes, has led to the exploitation of the finite resource of space. Space may formally be infinite, but that area of it which can be used for satellites is limited if users are to avoid the problems of radio interference in the signals to and from the satellites (Marsh, 1985, p. 31). In the rush to utilise the available area, less-developed nations have been effectively disbarred from using it (Jayaweera, 1987). As early as 1976, a gathering in Bogota of representatives from less-developed nations complained that the distribution of places by the International Frequency Registration Board was biased in favour of the developed nations (Marsh, 1985, pp. 60–1). The problem for states who are left out of such arrangements is that it is hard for them to later reclaim the space lost to them. International governing bodies are notoriously weak because they lack the resources and legitimacy which would allow them to enforce any given policy, at least one that ran against the wishes of the dominant states. Inequality results, therefore, from restrictive access to a technology.

Inequality may also result from the way the costs of a technology are distributed. We may all benefit from the provision of electricity, but the dangers and damage incurred in its generation are not borne in the same ratio. Those who live near or work in nuclear power stations face a higher risk of contracting, say, leukaemia than members of the public located further away. The same people are also more vulnerable to any accident in the power station. Similar costs fall upon those who live near nuclear waste dumps. These are just specific examples of the general phenomenon whereby the use of a technology imposes costs unequally. Most of us get free rides on the backs of others. As we drive down motorways or travel in planes, it is too easy to forget those who live by these roads or under flight paths. Such people will not experience the technology of travel in the same way as the traveller.

From some political perspectives, the fact that any individual has to bear these kind of costs may be sufficient argument for eliminating the technology altogether. Others, though, may be less inclined to adopt so drastic a solution. They accept the need for the technology and therefore the need for inequities in shouldering the cost. What they question is the way these costs are imposed. It is too often those without a political voice who are forced to carry the cost of the technology. It makes political sense, for instance, to route flights over poor areas where the prospects of protest are less.

The unequal effects of a technology are not, however, always distributed within a single political system. Acid rain is caused, in part, by the emissions from coal-fired power stations. The acid rain generated by British power stations does not fall most heavily upon British forests; it is other countries who suffer the ill effects of British technology. In the same way, the explosion at the Chernobyl nuclear reactor caused damage to people, animals and crops far beyond the Soviet border. Whether the costs of a technology stay within a nation's borders or whether they are passed on to neighbouring countries, the political problem remains. Whatever the benefits of the technology, the disadvantages are not shared

equally and do not fall to those who are either the main benefici-
aries or those responsible for its management.

CHOICES

Domestic technology seems, on the surface, to represent unalloyed
gains for the human condition, and yet it can be viewed as a
decidedly mixed blessing. Although it might seem that the introduction of the washing machine, dishwasher and their like, simply serve
to free people of domestic drudgery, there is a danger that the
advertiser's rhetoric might be confused with reality. A 'labour
saving' device may appear to provide more opportunity for other
activities, and thereby to extend the choice available to individuals,
but is this what really happens?

There is no doubting the popularity of domestic technology. In
Britain, over 90 per cent of homes have a washing machine. And it
is not difficult to see why it is so popular. It does save labour. It
eliminates the effort of hand washing and putting the clothes
through the mangle. But there are costs as well as benefits. The
first of these is financial. While washing machines relieve people of
domestic drudgery, they increase the financial costs to their users.
This burden is, of course, exacerbated by the element of planned
obsolescence incorporated into the technology. Apart from the
private financial costs of the washing machine, there are public
losses incurred. Ursula Huws (1988) argues that the private
provision of services (such as laundering) erodes demand and
support for public provision, the absence of which hits the poor
hardest. Privately based technology also, says Huws, eats into
people's sense of collectivity or community. The technology works
to privatise their world. Finally, despite the promise of a saving in
labour, domestic technology may in fact increase the workload of
women, who are expected to achieve higher levels of 'domestic
productivity'. As Bereano and others point out:

> The time women spend on housework has not declined very
> significantly in the last 50 years, despite the increased availability
> of modern 'conveniences'. Rather, new household technologies
> have helped maintain the number of hours required for house-
> hold tasks, causing technological unemployment among women

domestics, seamstresses and laundresses (Bereano *et al.*, 1985, p. 162).

Real labour-saving comes in quite a different form. As Huws observes of the rich: 'you don't catch them up ladders using the latest Black and Decker drill with sander attachment' (Huws, 1988, p. 6). They hire someone else to use the technology.

The point is that the costs and benefits of a technology cannot be measured simply in terms of costs or efficiencies as they relate to a narrowly defined function:

> the prime gains achieved through the application of technology in the home do not appear to have gone to women. While technologies may have decreased the physical effort of housework, they have not significantly reduced the time or psychological burden involved, changed the household division of labour or released women to enter the paid labour force (Bereano *et al.*, 1985, p. 180).

Ruth Schwartz Cowan draws a similar conclusion in her study of the effect of domestic technology on the home. Where the introduction of technology in industry leads to greater specialisation, in the home it creates the expectation that the 'housewife' will be 'responsible for every aspect of life in her household' (Cowan, 1985, p. 197). This is not a measure of the technology alone. As Rothschild (1983) suggests, while domestic technology presently tends to reinforce women's oppression, it also may contain the potential for their liberation. Although domestic technology may limit choice, it has the potential to expand it.

This ambiguity in the effect technology has upon choices can be witnessed in other areas. The increasing use of technology by the police has profoundly changed the way policing is done; it has also altered the way policing is perceived by officers and public alike. This point is elegantly made by Mike Stephens:

> One simple mechanical device, the car, transformed much of policing when it was introduced . . . in the 1960s. Both the police and public later came to acknowledge that Panda car patrols had not been a great success since they created barriers to the constructive and more congenial relationships that formerly the police had enjoyed with the public. Many police officers

began to see 'real' policing as the type of activity carried out by car patrols; the previous general sense of satisfaction with walking a beat waned and became of lower status, while the car offered a warm, dry environment, and an opportunity for excitement. The hedonism associated with vehicular patrolling replaced the more stolid world of 'easing behaviour' and familiarity with a particular beat (Stephens, 1988, p. 60).

The choices available to the police and the practice of policing were changed by technology. A similar phenomenon emerged with the introduction of the Police National Computer (PNC). To justify the high cost of the PNC it had to be heavily used, which led to pressure on police officers to meet a quota of spot checks on, for example, cars and their drivers. This added to the workload of the officers, but not to their effectiveness. For the motorist, the effect of the policy was experienced largely as an inconvenience or as an unwarranted intrusion (BSSRS, 1985, p. 11).

The above examples of domestic and police technology make clear a general proposition about the effects of technology. While it may be the case that the introduction of technology changes and extends the choices offered to an individual, the shift in opportunities is rarely as simple as it might appear. There are hidden costs to the new possibilities and easily overlooked benefits in the previous arrangements.

The introduction of technology changes the array of choices in the context of a set of social pressures and expectations which themselves may foreclose as many options as are opened up. There are no easy generalisations to be made about how technology affects the choices available to groups and individuals, except to say that each new technology changes the pattern of possibilities. It has, for example, been observed that in 1990 fewer children walked to school than they did in the 1970s. In 1971, 80 per cent of children walked to school, by 1990 the figure had dropped to nine per cent. The reason given was fear of the danger of traffic (Hillman *et al.*, 1991). In short, the freedom provided by transport has been matched by a closing of other options.

Establishing the nature and character of the new opportunities, and comparing them with previous arrangements, cannot be achieved through a reading of the makers' specifications. Incorporated in the comparison are judgements about features of the technology which cannot be established by design and function alone. There are, for instance, arguments about the 'value' of visits to the launderette. Value cannot be assessed purely in monetary terms; effects on the sense of community and levels of employment have to be taken into account.

CONTROL AND PARTICIPATION

Recently it has become common to talk of computer software (and other aspects of technology) as 'user friendly'. The implication is that some forms of technology allow more control to the user, while others deny such control. Technology which is intended to reduce physical effort by the user is also claimed to provide greater control or 'flexibility'. Think of the dishwashers with their ever-increasing number of 'programmes' or cars with their growing number of 'additional features'. Whatever the particular example, the suggestion is that the effect of a technology can be measured in the degree of control it allows to those who use it or work with it.

It is not, however, always the case that new technology increases control and participation. It sometimes decreases both. Harry Braverman (1974) claimed that the introduction of computer-based technology served to 'proletarianize' the white collar worker. Their tasks were standardised, their autonomy diminished. But, as Braverman points out, not everyone associated with this form of automation experienced a diminution of control. With the routinisation of certain work practices, the power of senior management increases because they are able to use the new technology to monitor the work of those below them (Barker and Downing, 1980).

Similar claims have been made about the development of medical technology. Though the ostensible purpose of, say, foetal screening has been to reduce the risks to women, the actual effect of the technology has been to diminish the control allowed to patients. At one level, technology may reduce the opportunity to say what you want and get what you need. Jennett writes of how new medical technology may remove 'the comfort of contact, the time for talk' (Jennett, 1986, pp. 32). The technology becomes an unwarranted intrusion, being used inappropriately:

in the labour ward and delivery room women have objected, questioning the need for routine monitoring when they are involved in a physiological rather than a pathological process, and objecting to the restriction on movement that some such monitoring involves (ibid., p. 31).

The technology, rather than helping, hinders the patient's control.

Of course, not all patients react in the same way. Retaining control is not a universal desire. Some patients, for example, *prefer* to be subject to technology. They do not feel they are being properly looked after unless they are being scrutinised by some element of high technology – 'simple blood tests and ordinary X-rays are now old hat' (ibid., p. 31). Acutely ill patients, for instance, may welcome high technology.

Whether or not patients embrace new medical technology, it is clear that its introduction can have the effect of reducing the control they have over their treatment. Interestingly, though, this reduction of control for the patient does not automatically lead to an increase in control for the doctor. The commitment of funds to the acquisition of new technology puts a considerable obligation on medical staff to use it, even where it might not be appropriate. Such an argument need not apply just to high technology. David Hellerstein recalls how he, as a young medical student, was present at an operation on the stomach of an old alcoholic who bled to death on the operating table. In the final stages of cirrhosis of the liver it is apparently common for uncontrollable bleeding to occur. The surgeon who performed the operation commented that once he made the first incision 'he knew he wouldn't be able to stop the bleeding', but, he added 'what choice did he have?' (Hellerstein, 1987, p. 142). Hellerstein's point was, first, that particular technologies may not be appropriate, and that commitment to them reduces control and flexibility; and second, the user may have little option but to employ the available technology.

Satellite television also demonstrates how a technology may affect the distribution of control. Because satellite broadcasting makes possible the reception of television pictures transmitted from outside the country's national borders, it means that governments have less control over the images and information which its citizens receive. It also increases the power of the media barons heading the new global communications corporations. Ted Turner, head of

Cable News Network (CNN) and Turner Broadcasting System, once remarked, 'We spread the information around in order for people to get along a little better. Governments could use a little help' (Clarke and Riddell, 1987, p. 11). Others take a less benign view, asking whether these media empires are always servants of democracy, and whether the world is a better place when people in Norwich, England, can know what the weather is like in Washington (Wallis and Baran, 1990). The same attenuation of control can occur with official data. 'The use of advanced telecommunications', reports Lyon (1988, p. 116), 'makes possible the processing and storing of commercial and even administrative data at a distance from where it is collected or from those it affects'. Shifts in the location of control are experienced as changes in access to it. Whether this is regarded as a force for good or evil does not change the fact that technology can be used to alter structures of control.

The effect of technology on control is partly determined by the context into which it is introduced. It can enhance the power of the already powerful, but equally, although this is rarer, it can shift the balance towards the less powerful. This double-edged quality to the effects of technology can be seen in fact-gathering. The increasing use of computerised data collection and collation brings, on the one hand, more detailed knowledge of society, but on the other, greater potential for the abuse of information (Campbell and Connor, 1986; Weeramantry, 1983). Worthy surveys can become malign surveillance. Much depends on the management and control of the data, on the criteria used in collection and on the accountability of the collectors.

The relationship which surrounds the technology is not, though, the only determinant of its effect. It is possible for the technology to change the nature of the relationship, to establish new functions and practices, which in turn generate new forms of participation and control, as we saw with the emergence of domestic technology. In addition to the context of the relationship which surrounds the technology, it is also important to observe that the design of the technology can influence the kind of control available to users. Some items of technology are designed to prevent access to the amateur repairer – through the use of non-standard locking screws, for example. There are no laws which decree that a technology be 'user friendly' or 'user unfriendly'. Design, like context, is important in establishing the effect a technology has on control and participation.

LIFE CHANCES

Design does not just affect the degree of control that people are able to exercise over their lives; it may also affect their prospects of enjoying valued goods. We referred earlier to the way in which the architecture of a building may discriminate against or in favour of certain groups. Similarly, people's life chances can be enhanced or damaged by virtue of the technology they (are forced to) use.

All technologies have some element of danger attached to them, but the risk involved can vary. The level of safety designed into airplanes is not matched in cars. And some technologies make very little attempt to ensure safety of any kind. In 1990, North Sea oil rig workers staged a one day strike. Their demand was for an improvement in safety on the rigs – two years earlier the Piper Alpha platform in the North Sea had caught fire, causing the death of 167 workers. Following the disaster, *The Observer* (10 July 1988) reported:

> Oil platforms are inherently dangerous and designed to far lower safety standards than would be permitted on land. Whitehall sources admitted yesterday that the structures are designed 'to very poor safety margins'. They are, for example, designed to withstand all but the worst wave likely to hit them in 100 years. This sounds good, until compared with nuclear power stations or hazardous chemical factories, which must be designed so that there is only a one in 100,000 or one in a million chance of a catastrophic accident from *all* causes.

In short, different life chances are built into different technologies. As a worker, your chances of suffering an accident are not the same for all technologies, and this is not an inherent function of the technology itself. It is also determined by the safety measures incorporated into the technology and its usage. The implication of this is that we should study with a degree of caution the accident rates attached to different technologies. Claims about the dangers of, say, crossing the road are not just statements about a natural state of affairs; they are about a particular set of humanly-ordered arrangements. There are many ways in which these dangers could be reduced (through attention to the design of cars and roads, and the regulations applied to them). Furthermore, the 'dangers' of a technology are not always evenly distributed. Some people are afforded more protection than others. Cyclists and pedestrians are relatively less protected from the dangers posed by the car than are other car drivers. And as we saw with the case of the Bhopal chemical explosion, the same technology is not subjected to the same regulations wherever it is introduced nor is the impact of disasters spread evenly. The effects of a technology can, therefore, be measured by the way in which users (and those who live or work in proximity to the technology) have their life chances influenced. Importantly, though, this impact is not a set feature of the technology itself, but of its context, design and the regulations governing its use.

Some of the most dramatic technological effects on life chances are promised by biotechnology. Its consequences are captured in the somewhat misleading label of 'genetic engineering'. Here, at least in theory, it is possible to manipulate the genetic code, and thereby alter certain characteristics of human development, eliminate certain inherited diseases and gain control over the reproductive process. US scientists have uncovered, for example, the cause of sickle cell anaemia. They discovered that the disease resulted from a mutancy in gene coding for the synthesis of haemoglobin. A single amino acid (one of 574) is wrongly substituted. This tiny variation is enough to kill the carriers in their teens. 'This lethal effect', writes Francis Crick (1990, p. 106), 'is produced by a tiny alteration in just one of the organism's many genes (we know now that it is due to a single base change). Essentially just two molecules are defective, one inherited from the father and one from the mother'. Once the breakthrough in understanding of the role and action of genes was made, and once the knowledge of geneticists and biochemists was combined, the technology of gene manipulation became possible, and with it the ability to control inherited characteristics. It is hard to think of a technology with a more potent effect upon people's life chances. The question as to whether the effect is beneficial or malign need not concern us here, but is interestingly discussed by the philosopher Jonathan Glover (1984, Ch. 2).

CULTURE AND EXPERIENCE

So far we have concentrated on the tangible effects of technology, the physical dangers it poses through the practical consequences of its introduction. But there are other ways in which technology can have an effect. For example it can influence people's perception of the world they live in. This may not put them at risk, but it can profoundly change the kind of people they are and the way they think and act.

An abiding concern of many countries, however rich and powerful they may happen to be, is the way in which their traditional way of life is being, or has been, eroded by the intrusion of outside cultures, which it is claimed, have replaced or destroyed the set of indigenous and valuable social codes and practices, which were integral to the life that people lived because they shaped the relationships and roles that gave meaning to their existence. Technology can drastically alter cultural formations. Consider the possible effects of satellite-based communications:

> With the arrival of direct broadcast satellites and the multi-beam high-grain satellites capable of sending as many as 40 pro-grammes directly into one's TV set, bypassing all terrestial control systems, the stage is now set, at least technologically, for the obliteration of the cultures of poor countries (Jayaweera, 1987, p. 206).

Two particular phenomena are held to have done the most damage to local cultures in recent times. They are tourism and mass culture. Both are products of, and are dependent upon, the development of technology. Mass tourism is the creature of the development of air transport, just as mass culture is only possible because of the evolution of mass communications technology. And, it is suggested, with these technologies and the practices they support, cultures can be destroyed or homogenised.

But in each case, though the technology is an important element in the process, there is a danger of exaggerating its impact. Tourism did not occur just because of the emergence of air travel, and the desire for communication predated the telegraph. This is not to deny that the character of both is dependent upon the technology that made them possible, but it is to deny that mass communications or tourism are simply determined by technology. The relationship is a complex one, in which cultural form and technology interact dialectically. To show how this happens, I want to focus upon the technology of popular culture.

Popular music is inextricably linked to the development of technology; indeed the possibility of popular music depended upon the technology. The sounds of pop are those of the human voice – the voice of everyday conversations and emotions: the whisper, the cry, the yelp and the shout. These cannot be carried to an audience without microphones and amplifiers. It is only with the technology that the intimate sounds and inflections of the voice can be heard. We are so used to microphones now that we do not notice them, but our response to music is in fact patterned by the technology. What we hear as 'real' is what the microphone picks up. Just as what we think of as 'genuine' handclaps are actually electronically sampled ones (Goodwin, 1990, p. 265).

It is not just the sounds of pop which depend upon the technology. Pop was only able to become 'popular' (that is, have a mass audience) through the creation of radio and record players, which were crucial for disseminating the sounds. Today, the global market for pop music is dependent upon the implementation of new forms of communication, from the video recorder to the satellite dish and the cable network. Equally, many of the stylistic innovations in the cultural form itself have been the result of the exploitation of new technology. The Beatles made use of multi-track recording; 'progressive' rock of the synthesiser; punk of cheap recording technology; and rap of sampling. In each case, access to the new technology provided not just an opportunity to improve existing methods (to make records with less background noise, or whatever), but also to extend the available range of techniques, and thereby to find new ways of making music.

Technology does not just affect the access to cultural resources, therefore; it also changes the nature of the cultural form itself. This is as true for its consumption as its production. Eisenberg writes about how 'in 1877 music began to become a thing':

The process took several decades, because the early phonograph allowed only a thumbnail impression of timbre and because Edison, who was partly deaf, turned his device to nonmusical and painfully unmusical uses. . . . In America the critical moment, the moment at which one might pinpoint the reification of music, was 1906. In that year the Victor company introduced the Victrola, the first phonograph designed as furniture, a console in 'piano-finished' mahogany that retailed at $200 (Eisenberg, 1988, p. 13).

Dave Laing argues that the introduction of the record-player also drastically altered the experience of listening to music: 'the replacement of an audio-visual event with a primarily audio one, sound without vision' (Laing, 1991, p. 7). Walter Benjamin (1970) made a similar point in his famous essay on 'The Work of Art in the Age of Mechanical Reproduction', written in the 1930s. Benjamin argues that the technical ability to reproduce art changes its nature. Before the possibility of mechanical reproduction, works of art existed only at the moment of performance or on the gallery wall. With mechanical reproduction, the work of art is no longer bound to a particular place and time. 'Technical reproduction', writes Benjamin (ibid., p. 222), 'can put the copy of the original into situations which would be out of reach of the original itself'. In losing uniqueness, the mechanically reproduced work of art also loses what Benjamin describes as its 'aura': 'the essence of all that is transmissable from its beginning, ranging from its substantive duration to its testimony to the history which it has experienced' (ibid., p. 223).

Benjamin was not simply observing a change in the means of making and reproducing art. He saw a fundamental change in human experience: 'the mode of human sense perception changes with humanity's entire mode of existence' (ibid., p. 224). The particular change being made is one which in which art is no longer founded upon ritual but upon politics. Earlier, non-reproducible art served to enhance rituals, through which art's authenticity was guaranteed. With reproduction, the claim to authenticity or uniqueness no longer makes sense; there is no ritualistic power to be claimed. The function of art has changed so that it has now become integrated into reality. 'The painter maintains in his work a natural distance from reality', wrote Benjamin (p. 235), 'the [film] cameraman penetrates deeply into its web'.

The change in the form and aesthetics of art is combined with the opportunity for mass participation. Film, unlike painting, is intended for collective appreciation. It also changes the quality of that appreciation:

> By close-ups of the things around us, by focusing on hidden details of familiar objects, by exploring commonplace milieus under the ingenious guidance of the camera, the film, on the one hand, extends our comprehension of the necessities which rule our lives; on the other hand, it manages to assure us of an immense and unexpected field of action (ibid., p. 238).

'The camera', Benjamin continues, 'introduces us to unconscious optics as does psychoanalysis to unconscious impulses' (p. 239).

Andrew Goodwin (1990) adds an intriguing twist to Benjamin's argument. The emergence of digital audio technology (CDs, digital audio tape, video-discs) further demystifies the 'aura' of art. Almost all recordings are now digital, as opposed to analogue, which means that there is no loss in quality during reproduction. This effectively means that 'everyone may purchase an "original"' (ibid., p. 259).

Benjamin's argument is not without its critics. His account overlooks the wider material context in which the technology of artistic reproduction is organised, and as a consequence it works with a form of technological determinism, by which the technology is the exclusive cause of the change in perception. But these criticisms, as Middleton points out (1990, pp. 68–9), need not detract from Benjamin's insights into the way that 'the new modes of production and reproduction would generate new kinds of perception'.

In this section, we have seen how technology can be said to affect the ways in which people perceive the world, and thereby change their reactions and behaviour. The technology of mass communications does not simply reproduce pre-existing modes of thought, it engenders the creation of completely new ones. Not all technologies will work in this way, but some will. A similar kind of argument could, for example, be made for the introduction of the clock and the formalisation of time that it made possible (Landes, 1983; Thompson, 1967).

SIDE EFFECTS

The World Commission on Environment and Development (1987, pp. 4–5) observed in its report, *Our Common Future*: 'A mainspring of economic growth is new technology, and while this technology offers the potential for slowing the dangerously rapid consumption of finite resources, it also entails high risks, including new forms of pollution and the introduction to the planet of new variations of life forms that could change evolutionary pathways'. This is what I mean by side effects.

While some of the effects discussed earlier in this chapter may pass unobserved, there are others which have attracted considerable attention, and have led to vehement opposition to technology. Pollution is the classic example of this. Damage to the environment is rarely the intention or purpose of a technology; it is often, though, a consequence. The point of this section, then, is to draw attention to the side effects of technology, effects which are not intrinsic features of the technology and its operation. It is worth observing in passing that though most such side effects are seen as detrimental, they need not necessarily be so; there is no logical reason why the side effects should not be beneficial. What is important is their unintended character.

Lead emission from car exhausts is one such side effect. Lead was introduced into petrol to prevent 'knocking' in the engine. Knocking is caused by premature ignition of the petrol pressurised in the cylinder and results in the engine running less efficiently. But having put lead in petrol, it inevitably becomes part of the waste products emitted from the exhaust pipe. Lead is a poison, and it can be especially damaging to young children. It attacks the brain, and can cause learning difficulties and behavioural problems. Medical research tends to confirm that car exhaust has been a cause of ill health in the young.

One problem with such side effects is the ability to demonstrate the relationship between cause and effect, precisely because the cause is not directly related to the intention or the purpose of the technology, and because the effect may be explicable by other means. It was exactly such problems that led the Lawther inquiry, commissioned by the British Department of Health and Social Security, to conclude in 1980 that no definite connection could be made between car exhaust and child ill-health. As it happened, this position was subsequently undermined and the government eventually agreed to reduce the level of lead in petrol.

Similar difficulties emerge with the effects of acid rain. Acid rain was first identified in the 1960s in Scandinavia. Researchers reported the death of large numbers of fish in Swedish lakes. By the 1980s, every European country was known to be experiencing the ill-effects of acid rain. The acid content of the rain was produced by sulphur dioxide emissions from power stations. The problem could not have been anticipated at the time at which the generation plants were being built.

One important feature of side effects like acid rain, or the fallout from Chernobyl, is that the effect is felt some way from its primary cause. Agricultural technology, in the form of nitrogen fertilisers, causes the pollution of water supplies that reach thousands of homes over a wide radius. Sometimes this can result in the death of the very young:

> Nitrates are converted by bacteria in the human gut into nitrites, related substances which can combine with the oxygen-carrying pigment in the blood. If this happens to an excessive degree, a condition called methaemoglobinaemia results in which the victim effectively suffocates from within as the blood cannot carry adequate oxygen for the body's needs (Price, 1984, pp. 79–80).

Once again, the side effects are not distributed evenly. This is also apparent in the experience of less-developed countries. Technologies, imported from the West, are imposed upon an environment and social structure which do not 'fit' the technology. The result is a more devastating form of pollution (Redclift, 1986).

A further feature of these side effects, which is already apparent but is worth emphasising, is that many individual causes may lie behind a very general effect. The erosion of the ozone layer and the greenhouse effect, for instance, are not the result of the activities of a single technology or a single user. The effect of carbon dioxide, chlorofluorocarbons and other emissions on the earth's protective ozone layer and atmosphere were discovered in the late 1970s. The discovery was made possible by the development of another technology, satellite surveillance, which provided previously unobtainable information on the state of the atmosphere. The implications of ozone erosion and global warming are only slowly being recognised, but there is no doubting their seriousness. As O'Riordan (1990) comments, 'The political challenge of the boosted greenhouse [effect] is that of the end of national sovereignty as we know it'. The extent and unpredictability of the side effects are precisely what makes a coherent political judgement or response so difficult. These problems are compounded by the intricacies of cause and effect, intention and responsibility which are also incurred.

CONCLUSION

This chapter has, for the most part, simply observed the possible effects of technology, drawing attention to the arguments and ambiguities to be discerned in its implementation. We have seen how the introduction or development of technology may have an impact on a range of issues that are central to the study of politics. Technology may affect the opportunities, power and experiences of people and institutions. It can have an effect upon thought and behaviour, and upon the means by which they may be controlled. At the same time, it has become apparent that these effects are created by a range of factors, which extend from the technology itself to its context. There are, therefore, no easy generalisations to be made about 'the effects of technology', not least because all effects are open to interpretation (whether of their cause or of their consequences). There is certainly a danger of concluding that the effects of technology are in a sense *determined*. This would be wrong. The point is to recognise the *potential* effects which may flow from any given technology. Many of them can be changed, the question is how? By altering the technology? Or its use? Or its context?

There are no simple answers because, as we have seen, not only are there many possible effects, there are also many different ways of interpreting them. Any attempt to judge and control a technology must, therefore, begin with a detailed analysis of its actual/potential effects. It then depends upon an examination of the causes of these effects. And finally it requires the exercise of judgement as to which effects are to be moderated or enhanced, and by what means. This process of analysis means considering a large number of possibilities. Bryan Jennett's discussion of medical technology demonstrates how it might proceed:

high technology can relieve the doctor and nurse of time-consuming chores, of longhand notes and charts, of filling forms and taking repeated observations by hand and eye. That should mean more time for talking to patients, for reassurance, for explanations to anticipate concern about the next close encounter of a technological kind (Jennett, 1986, p. 33).

What Jennett recommends is a close examination, within a wide remit, of the causes and effects of the deployment of medical technology. But it is one thing to say what ought to be done, it is another to do it.

6 Choosing Technologies

INTRODUCTION

Recognising the political effects of technology is one element in establishing democratic control over it. The other crucial element is determining how we should judge technology and its effects. How do we make a coherent choice between technologies? How do we measure malign effects against benign ones? Is there a rational, democratic way of choosing technologies? These questions are the concern of the rest of this book. Our attention, therefore, is shifting from the institutions and the circumstances which currently frame technological decisions, to the arguments about the political ideas and possible procedures to be applied to those decisions.

This chapter establishes the groundwork for exploring the political arguments around technology decisions. It does this by looking at the sort of questions which any such debate must address and answer. We need to begin by asking what sort of difficulties confront us, whatever our ideological predisposition, when deciding about technology. Of course, not all forms of technology pose the same problems; and the problems posed are rarely – if ever – exclusive to a single technology or technology in general. This section, therefore, just identifies some of the issues which emerge whenever decisions about technology are made. They can be fitted into three broad categories. These are, first, informational: how do we know what the technology will do and what effects it will have? Secondly, they are judgemental: how should we evaluate the possible outcomes, and what criteria should be included in the assessment? And finally, there are the institutional problems: how are decisions to be reached and who is to take them? In keeping with these broad categories, the chapter is divided into three sections: (1) the character of decisions about technology; (2) the assessment of technology decisions; (3) the institutions of risk and technology assessment.

THE CHARACTER OF DECISIONS ABOUT TECHNOLOGY

What sort of decisions are involved in choosing technologies? What particular difficulties do they represent? Unless we answer such questions, we are in no position to assess the political process which makes these choices. Here I concentrate on three elements of technology decisions: bias, uncertainty and unknowability. As Roger Williams (1989, p. 164) observes, 'Policies towards technology can rarely escape uncertainty and risk and the unknowns are many'. My account is selective; there are many other dimensions and many fuller discussions (see, for example, Fischhoff *et al.* [1983] and Roberts and Weale [1991]).

Biased choices

It might be imagined that choosing between technologies is straightforward. Either you adopt it or you reject it. The choice is rarely this simple. Choices do not divide so neatly that the merits of each can be equated easily. David Collingridge (1981, pp. 85–102) demonstrates this asymmetry in the case of the choice between a power station programme and a conservation policy. Consider the position of a government faced with these alternatives. If the former was chosen, the worst result possible would be failure to fill the energy gap; an energy shortfall would presage economic collapse and a political crisis. Therefore the best solution would be the relatively cheap conservation policy. There would be no need to build expensive power stations; the burden would be carried by the individual consumer. There is, however, a catch with this solution. Although a conservation policy would serve the government's interests, it would be very difficult for it to be put into effect as the government would have to depend on the cooperation of the consumers. Given the cost of failure, the government would be reluctant to rest its political future on the goodwill of its citizens, especially if there were an alternative course. And there is: to build power stations. This would have the advantage of producing definite results and of being within the remit of government. It would have the disadvantage of being costly, but the costs would be small compared to those which would be incurred by an energy gap. The logic of this argument is that 'conservation' and 'power stations' are not equivalent. There is a bias in favour of the latter, even though it is not the ideal solution. Governments, as Collingridge shows, will tend to hedge their bets by going for power stations (and in the process over-supplying electricity) because it provides the only sure way of avoiding the worst possible outcome.

O'Riordan suggests that a similar thinking presently affects calculations about the appropriate response to global warming:

> the problem for the politician is simple to express. The science is alarming but not proven. There appears to be time to be sure, so it is comforting to delay while the modelling and predictions improve. The economic consequences of restructuring energy use, reducing or heavily pricing automobile transport, cutting back on livestock and rice-based agriculture (important sources of methane) are unknown, but could be devastating (O'Riordan, 1990, p. 14).

The same, cautious logic informs the decision (discussed in Chapter 3) of companies who choose to commit themselves to government defence contracts rather than civilian projects (which might, but only might, generate greater income). It also characterises governments who remain committed to a policy programme because of the resources they have already committed to it. Tony Benn, then at the Ministry of Technology, records in his diary about how he had to choose between different kinds of nuclear reactor for Britain's nuclear energy programme. There was pressure on him to move away from the design of reactor produced by the United Kingdom Atomic Energy Authority (UKAEA). 'One could see all this huge investment in nuclear research absolutely going down the drain. It was a shock to hear it because we had spent about £900 million and the question was whether any of these systems are as good as we thought' (Benn, 1988, p. 58). In the end, both Benn and the ministers who succeeded him remained loyal to the UKAEA designs (Greenaway *et al.*, 1992).

There are two types of factor which contribute to the bias that affects decison-making. Firstly there are *exogenous* factors which make one solution more favourable than another. For example, opting for conservation rather than building power stations would be viable only in circumstances where a government had the capacity or the will to enforce conservation. The bias here is a function of the political structure that surrounds the choice. Biases of this kind can, in theory, be corrected by reforms of the political structure. But this solution, even if it were acceptable, is not viable if the bias is *indigenous* to the technological choice itself, where no amount of political reform will alter the basic character of the choice involved. This would seem to be the case with the choice between medical treatments, where one requires surgery and the other requires the patient to adopt a personal regimen of some kind. AIDS illustrated this dilemma. The choice here is between developing a cure for the disease and teaching the practice of safe sex. In this case, of course, the choice is not a stark one and it is possible to combine the two options, but it is also clear that only one (the cure) guarantees success. Choosing technologies, in other words, rarely involves comparing equivalent options. Each one builds upon different assumptions and offers different kinds of result.

Uncertainty

The reason some choices are biased is that they vary in the degree of certainty they offer, and uncertainty is a major force where decisions about technology are concerned. Writing about the use of high technology in medicine, Jennett observes: 'There is no other way to judge the wisdom of a value judgement than by awaiting the outcome of decisions based on it and then making a value judgement about this' (Jennett, 1986, p. 257). A judgement has to be made before the results can be assessed. In other words, decisions about the implementation of high technology involve the need to exercise judgement in the face of uncertainty. This obviously applies to new technology, but it also has relevance to the new applications of existing technologies.

If a decision involves assessing the costs and benefits of a technology, and if the decision has to be taken before the technology has been introduced, then it is unlikely that we can predict exactly what effects the technology will have. In the case of technological innovation, the uncertainty extends to an even wider set of issues – from its completion date to its efficiency. Britain's nuclear power programme has amply illustrated these types of uncertainty. Predictions about the cost of generating electricity and about the time when the reactors would come on stream were notoriously inaccurate (Williams, 1980).

Even a technology that has been in constant use is subject to uncertainties. The life expectancy and reliability of a technology will depend upon many factors. There is the uncertainty engendered by the various uses to which a technology may be put. (Compare the treatment – and longevity – of cars.) This may be compounded by external factors which affect performance. Shifts in the world price of oil, for example, can radically affect the running cost of some technologies and thereby determine their viability (the domestic central heating system is just one obvious example). While such price fluctuations may be expected, their character is inherently uncertain.

To point to the various sources of uncertainty is not, however, to conclude that ignorance rules. Different types of uncertainty can be distinguished, depending on their susceptibility to measurement and control. Following the sinking in June 1987 of the cross-channel ferry, the 'Herald of Free Enterprise', in which 193 people died, there was a protracted legal case about who was responsible. The accident was caused by a failure to close the bow doors to the car decks – in rough seas just outside port, water entered through the doors and the ferry sank. The management of Townsend Thorensen, who ran the ship, were accused of having failed to anticipate the event which led to the sinking. They should, it was claimed, have provided a warning system by which the captain could ascertain from the bridge that the bow doors had been shut. The force of this accusation was given additional weight by the argument that the company had been told repeatedly of the potential hazard, but had chosen to take no action. For the critics of the company, the uncertainty represented by the chain of events which led to the accident should have been anticipated and prevented – the uncertainties could have been guarded against. Opposed to this argument was the view that the management could not reasonably be expected to take responsibility for all their employees' actions. Resolving such questions is, of course, vital to the process of obtaining legal redress.

Uncertainties are intrinsic to the development of technology. To demand they be eliminated or guarded against might, in effect, prevent the development of further technology. The cost, in time and money, might have the overall effect of making the technology economically or technically unviable. This might, of course, be the right result in some cases. It might, for example, be used to prohibit any attempt at the manipulation of human genes or to prevent the further development of the fast breeder reactor, on the grounds that both the uncertainties and the size of the possible outcomes make the costs of any mistake too high. But such arguments cannot apply to all technologies. There is a point at which it is judged to be 'safe enough', when a sufficient range of possible outcomes has been considered and where the technology remains technically and economically viable. In short, a judgement has to be made about the uncertainties confronted.

In observing the limits to our ability to anticipate events and the irremedial character of uncertainty, we are not, however, committed to the conclusion that we must be quite agnostic about the future and, less abstractly, that the managers of the 'Herald of Free Enterprise' were blameless. As we have already suggested the *type* of uncertainties can vary between technologies – the dangers threatened by a car are greater than those threatened by a children's bicycle. And within a given technology, we can identify priorities in the uncertainties, so that some receive more attention than others.

Equally, there are varying degrees to the difficulty of anticipating different uncertainties. It is obviously easier to identify short-term uncertainties, rather than ones which exist in the long term. This may account for the fact that far more attention has been devoted to the uncertainties surrounding the daily operation of nuclear reactors than to their fate when they are decommissioned. It may explain this disparity; it does not necessarily justify it.

Given these varieties of uncertainty, there are ways in which both the type and the anticipation of them can be affected by the introduction of appropriate procedures and agencies. The lesson of the 'Herald of Free Enterprise', for instance, might be to evolve a consultation process which would give due weight to the advice of those with specialised knowledge of potential dangers. As it is, most states employ some system of technology vetting, whether it is the United States Office of Technology Assessment or Britain's Health and Safety Executive. My concern here, though, has been to draw attention to the problems with which any such organisation has to deal.

Unknowability

Uncertainty is a subset of what can be called unknowability. Uncertainty is caused by the fact that some things cannot be known. Many of the questions we might like to answer before taking a decision may be unanswerable in at least one of two ways. First, they may involve questions of judgement – at the simplest level, about what is to count as a 'good' or 'bad' outcome – which can never be resolved absolutely. One reason for this is that people will, whatever their expertise, disagree as to what issue is at stake. Take, for instance, the problem of traffic congestion. From one perspective this problem arises because cars cannot travel at the speed for which they have been designed and roads do not allow for the efficient use of cars. From a different perspective, the problem arises from the fact that people are forced to resort to cars in the absence of an adequate, non-polluting public transport system. These two perspectives agree neither about the problem nor its solution, and no attempt at reconciliation will resolve it because there is no knowable answer.

A second kind of unknowability is technical. There is, for example, no way of knowing exactly the effects of a modern nuclear explosion. Our knowledge depends on the evidence of the bombs dropped on Hiroshima and Nagasaki, and on theoretical extrapolation. We cannot be sure whether or not a 'nuclear winter' will follow a nuclear exchange; and we know even less about what it will be like. The only point at which these things can be known is after the event has happened – which is too late. More mundanely, there is the decision not to include smoke hoods as standard safety devices in airplanes. The aviation authorities decided against the hoods on the grounds that for every one life saved by them, eight would be lost in the delay that might result in the evacuation of the plane (*Independent*, 25 April and 27 May 1991). Their decision was based on a prediction of how people would behave. Prediction is necessarily an inexact science.

All lists of the possible failures of a technology are inevitably insufficient. We cannot, for example, anticipate all the mistakes an operator can make; nor can we identify all the contexts in which a technology will be used (you might call this the 'blunt instrument' phenomenon: almost anything can become a murder weapon in the wrong hands). Even if we can add to the list of possible consequences of a particular technology, we could never furnish a complete list. If every moment in time is in some sense unique, in that it represents a set of circumstances which have never previously existed in that particular form, and if it is impossible to provide a complete list of any one set of circumstances, then we cannot anticipate all possible outcomes. Some consequences are bound to be beyond anticipation.

Unknowability is different from uncertainty because it does not allow for comparative analysis, and has no analogy in existing experience. Uncertainty refers to the difficulty of determining the prospect of a particular outcome. Unknowability refers to the problem of knowing what outcomes are possible. The two are similar in one important respect; they force decision-makers to use *judgement*. In the case of uncertainty, a judgement has to be made about committing resources in order to reduce ignorance. In the case of unknowability, judgement is exercised in determining the wisdom of pursuing a particular technology, given that we cannot know what its consequences may be.

THE ASSESSMENT OF TECHNOLOGY DECISIONS

Given the biases, uncertainty and unknowability involved in decisions about technology, we need to ask how we begin to systematise those decisions so as to reach the best possible outcome. One task is to assess the risk of the possible results (risk assessment); the other is to assess the technology itself (technology assessment).

Risk assessment

The exercise of judgement over uncertainty and unknowability finds its formal representation in the business of risk assessment. Given that the outcomes of a technology are inherently grounded in speculation, the decision to proceed with it must be based on a calculation as to whether the probable outcomes are favourable or disadvantageous. Is the technology a good or a bad risk?

There are two stages in risk assessment. Firstly, there is the question of identifying the risks involved – what can go wrong? And what chance is there of this happening? The second stage involves deciding what risks are worth taking. They both include elements of uncertainty and unknowability, and they both require an act of judgement. They are also inescapable features of decisions about technology. There is a vast literature on risk assessment, and

I shall make no pretence to summarise it here. Instead, I want to draw attention to some of the problems entailed in it – the risks of risk assessment.

People are, on the whole, notoriously bad judges of risks. Part of the reason for this, argues Paulos (1990), is that they are unable to think sensibly about numbers. Whenever the threat of terrorist attacks on aircraft is given public attention, people rush to cancel their seats. During the 1991 Gulf War, for example, many US showbiz stars cancelled trips to Europe for fear of Iraqi-sponsored terrorism. But even at times of heightened tension, the actual risk of being the victim of such an attack is minuscule. There are many other more dangerous activities which the same people enagage in daily without a second thought – travelling by car, for instance.

Numbers are not the only cause of problems in the popular assessment of risks. Words can create difficulties too. Fischhoff et al. (1983, pp. 25–7) report how people judge the same risk differently, depending on how it is described. They show how different responses can be elicited, depending on whether intervie-wees are told how many *deaths are caused* or *how many lives are saved*. The options are assessed according to the way they are phrased, not according to the facts they refer to.

But even when risk assessment is not vulnerable to the problems people have with numbers or the associations they attribute to words, values and judgements are integral parts of the process. Here the problems are manifold. Different values may be put upon the same outcomes; or the results may be given different priorities; or there may be differences over what the results are. Any technology which some see as threatening the natural environment is full of such questions. How is the loss of, say, a hedgerow to be evaluated, and how is it to be compared to gains in economic or technical efficiency? In examining such questions, there is a need to be aware of where the values being used emanate from. A desire to protect wildlife may be born of a love of nature; it may equally derive from a self-interested desire to preserve the commercial value of the site. Are judgements of risk based on the protection of self-interest to be taken more or less seriously than disinterested concern over the natural environment? More technically, there are difficulties in determining the possibility of very unlikely events which have potentially disastrous consequences. A melt-down in a nuclear power station or a passenger plane landing on an urban centre are examples of these. Even if a value can be put upon such risks, there is still the problem of comparing it with the risk of more determinant, but less serious, outcomes.

Judgements need not only be confined to the results of a technology. There are intense arguments over the way the cause of a risk is assessed. The nuclear lobby likes to argue that the risk of cancer being incurred through radioactive emission from power stations is minute when compared to our normal daily exposure (from watches, television sets, rocks, and so on). The anti-nuclear movement responds by arguing that there is an important difference between background radiation and humanly created radiation, and between risks that are chosen and those that are imposed. Whichever side anyone takes in such disputes, the argument is inescapably tied to judgements, which are themselves shaped by political and moral values. As O'Riordan (1983) comments, 'Risk has a moral element: it is a manifestation of how society through its diverse groupings passes judgements upon itself and its institutions'.

To point to such problems (and there are many more) is not to claim that risk assessment is either arbitrary or impossible. It is merely to draw attention to the fact that any system which seeks to decide about technology has to confront – and resolve – them. What this review reveals is that: (1) the system must acknowledge the need for judgements, and (2) there is no obvious answer as to how the judgements should be made and by whom. We have, for instance, seen how popular participation may be vulnerable to extraneous influences. Competing solutions will be considered later, but for the moment we need to look further at the difficulties entailed in decisions about technology.

Technology assessment

Underpinning risk assessment is the attempt to assess the effectiveness, safety, reliability, or whatever, of the technologies concerned. Ham and Jennett (1987, p. 2) identify three basic tasks that any technology assessment has to fulfil. It has to establish how *effectively* the technology will meet its given goal; how *efficiently* this will be done; and whether it constitutes a *socially acceptable* method. In trying to determine each, a number of problems will be encountered; I shall focus on three of these.

Lead times

The time taken from inception to the implementation of a technology can be very long or very short. Both cause problems. Short lead times leave little opportunity for assessment. Governments are forced into a position where they are coping with, rather than controlling, the technology. Both biotechnology and information technology have been experienced like this. Although the underlying science developed gradually, once it had been converted into a commercial form its implementation occurred rapidly and spread widely.

Long lead times are no easier to manage, and indeed may pose even greater problems for assessment. With some high technologies – nuclear power or major road building projects, for instance – a very lengthy development period is involved and decades may elapse between the initial decision and completion. Two problems derive from this. Firstly, assessments conducted in the early stages of development require anticipation of the distant future, with all the uncertainties that this inevitably entails. Kirkegaard once remarked that 'Life can be understood only by looking backwards, even if it has to be lived looking forwards.' (Magris, 1990, p. 41). What is true for life is also true for technology assessment! The second problem caused by long lead times is that when the uses and operation of such a technology become clear, the opportunities for changing (or halting) the development become very limited. In this sense, assessment is almost redundant, except as an exercise in anticipation rather than direct control.

Costs of assessment

The actual business of assessing a technology can itself add to the problem. Most obviously, technical assessment costs money, and this money has to be added into the balance of costs and benefits which are used in weighing up the technology. Assessing medical technology, for example, entails clinical trials and economic costings. These tests have to be included in the costs of the technology and have to compete for the scarce resources within a health service. The result is that the original choice between various medical strategies and their associated technologies will be changed by the costs entailed in testing each (Jennett, 1986, pp. 257–8). A cheap technology may require expensive assessment, or vice versa.

A classic case of the impact of assessment on the thing being assessed was the Sizewell B inquiry which took place in Britain in the early 1980s. The Central Electricity Generating Board (CEGB) proposed to introduce a new design of nuclear reactor. A public inquiry was necessary, and this turned out to be a long and expensive affair. Providing detailed evidence, briefing lawyers, maintaining an active public relations campaign and delaying the building work, all imposed a considerable financial burden (about £10 million) on the CEGB. These costs meant that estimates of the economics of the power station were changing constantly (O'Riordan *et al.*, 1988). These expenses were over and above the £7 million that was spent by the government-sponsored Nuclear Installations Inspectorate on assessing the original design. The longer the inquiry went on the more the reactor cost. (Like the protracted legal wrangle over an inheritance in Charles Dickens' *Bleak House*, we might suppose that if the inquiry had run on for many more years, its costs would have exhausted the resources available for building it.) It is not perhaps surprising that the government reacted to the Sizewell B inquiry by proposing to cut down the scope of such exercises in the future.

Sources of information

Inquiries do, however, provide an almost unique opportunity for information to be collated and placed in the public realm. In Britain the Sizewell B inquiry added greatly to knowledge about nuclear power (O'Riordan *et al.*, 1988). The House of Commons Social Services Select Committee provided a forum for drawing together disparate knowledge and advice on AIDS. The Warnock inquiry into *in vitro* fertilisation provided a clear public statement of the key issues and the state of the microbiological art. In the USA, the Office of Technology Assessment provides an important source of information on new technologies.

But whilst inquiries do provide an opportunity to collect information on a particular topic, the source of the information and its character may pose additional problems. First, time and money are required to collect and present information, and the access to this information is not equally distributed amongst all who have an interest. At one level, this may simply be a matter of finances. In the argument over the siting of a road, a ministry of transport can command far greater resources than local residents. No public money is provided for protestors. When the inspector in charge of the Sizewell B Inquiry asked the government for funds for the objectors, his request was refused because, said the Department of Energy, it would set an 'expensive precedent', and besides the inquiry system provided adequate protection for the public interest (Kemp *et al.*, 1984, p. 484). At another level, access to information may be easier for some interests than others. For example, in a dispute over road siting a ministry of transport can call upon the services of road engineers and planners, but protesters may have access neither to experts of this type, nor to other forms of expertise (which the ministry might not wish to be included) that relates to, say, noise or exhaust pollution. Secondly (and relatedly), there is a tendency in all inquiries to apply a selective definition of 'relevance' or 'authority' to the information received. This can be practised through the language and ethos of the inquiry, often borrowed from those of the courtroom. It can find direct expression in the use of lawyers to represent the various interests and to cross-examine the witnesses. As a result, certain arguments do better than others. In particular, an argument grounded in 'scientific expertise' carries more weight than one based on 'political judgement'. One witness to the Windscale inquiry in Britain (about the siting of a nuclear waste reprocessing plant) remarked:

> If one were to say 'I oppose nuclear power because it threatens the very stuff of life', one would be branded as emotional and incapable of rational argument . . . there is far more scope for interaction if one says 'I oppose nuclear power because the disposal of long-lived actinides to geological formations of unproven permeability or long-term structural stability cannot guarantee against radiological hazards due to concentration in food chains' (Breach, 1978, p. 171).

Indeed, the overall effect of an inquiry process based on a legal format tends to remove politics altogether, and may even lead to it operating as just a means of refining the arguments of the dominant actors (Kemp *et al.*, 1984). New evidence or issues do not so much undermine the case being made as provide fresh material to be incorporated within it.

While all or none of these problems may be encountered in the business of taking a decision about technology, it remains the case that all information is partial, and decision-making has to acknowledge this. Taking decisions about technology means confronting the problems posed by both risk and technology assessment, each of which are subject to the effects which result from value judgements made under conditions of ignorance. Institutions charged with such decisions are intended to provide a systematic method for coping with these problems.

THE INSTITUTIONS OF ASSESSMENT

One central message has become clear from the discussion of technological decisions. It is that all such decisions require judgements. This is true of all political decisions, but in the case of technology the claims of the politician or the public are challenged by at least one other source of authority – the technical expert. Below we review the competing claims of those who believe that decisions should be 'left to the experts', and those who would recommend some form of public participation.

Expertise

One obvious solution to the problem of taking complex, technical decisions is to give the job to experts, people with specialised knowledge of either the technology itself or of the analysis of risk. The assumption is that experts provide a judgement on a technology which is not clouded by ignorance or prejudice. But is this a reasonable claim? Expertise is an expression of the increased division of labour, itself a product of the development of technology. Technical change involves the constant extension and refinement of the division of labour through the application of abstract scientific knowledge to each aspect of a product or process. For each stage there are specialists. In this proliferation of expertise a number of problems emerge. The most obvious is that there is no one voice to speak for a technology. This means that for each technology there are competing claims to knowledge. Though the experts may be talking about the same technology, they do not do so from the same perspective. Engineers work within a different paradigm than do, say, risk analysts, and their insights cannot simply be amalgamated to provide a single perspective on a single object. Indeed, it is difficult to see how an overall perspective can be established. Each expert knows a great deal about their particular speciality, but they know much less about the project as a whole. Yearly says of specialism within science:

> As scientific knowledge becomes increasingly specialised, scientists are less and less able to pass judgement on neighbouring areas of science on the basis of internal criteria. Someone with knowledge of one field will almost certainly not be competent to pass internal judgements on adjacent disciplines (Yearly, 1988, p. 81).

A similar argument can be made about technologists. Indeed, it can be argued that with complex technology no overview is *possible*. Without this, though, there is no vantage point from which to judge the value of the technology as a whole. It is feasible only to look at different aspects of the process. This problem is nicely captured by a vice-president of research at Eastman Kodak:

> The best person to decide what research work shall be done is the man (*sic*) doing the research, and the next best person is the head of the department, who knows all about the subject and the work; after that you leave the field of the best people and start on increasingly worse groups, the first being the research director, who is probably wrong more than half the time; and then a committee, which is wrong most of the time; and finally, a committee of vice-presidents of the company, which is wrong all the time (quoted in Zuckerman, 1971, p. 112).

There is a further feature of the partiality of this expertise. Experts have interests, albeit not ones that fit neatly into the ideological spectrum, and these interests are closely tied to and shaped by the technology. The more complex or vast the technology, the greater the degree of specialism and the more specific that expertise is to the process or product. A threat to the process or product is a threat to the expert. For experts in nuclear power engineering, the end of the nuclear power programme may mean an end to their career. Their skills will be difficult to transfer. Insofar as experts become locked into technology in this way, their judgement of it cannot be 'independent'. A former British government chief scientist commented critically on the lack of impartial advice on nuclear matters. 'It needs no crystal ball', he wrote to *The Times*, 'to see what advice UKAEA will give civil servants' (quoted in Bacon and Valentine, 1981, p. 78).

Finally, it is important to recognise that 'expertise' is socially created. What counts as authoritative knowledge on a subject is determined by the operation of values and judgements. We have only to think of the relative weight given to the expertise of, say, scientists, doctors, social scientists and priests in different places at different times. This is not to claim that there are not good reasons for preferring one expert over another. Rather it is to draw attention to the fact that, in assessing technology, there is a need to be aware of the source, status and interests of expertise.

Public participation

The problems posed by expertise sometimes translate into a decision to rely upon the democracy of popular choice, rather than elite judgement. If all experts are partial, if their advice is biased, then there has to some way of checking or countering their claims to authoritative knowledge. Typically, though, expertise is framed by a professional organisation which, while operating an internally regulated professional code, may not be subject to any public form of regulation. In such circumstances, it might seem reasonable to include the advice of experts with all other forms of partial political advice, most obviously with that provided by interest groups. In short, decisions about technology should be decided by the kind of political participation that is recommended by pluralist accounts of democracy, in which interest groups compete for policy outcomes (for example, see Dahl, 1956). This kind of argument is reviewed in more detail in the next section. Here we need only to make a couple of points about whether pluralism can provide an adequate forum for views on a technology.

There are, as the long-running debate about pluralism reminds us, many reasons to doubt the adequacy of interest groups (see Held, 1987). Apart from the familiar barriers to access formed by money and language (it is no coincidence that interest groups are predominantly middle class), there are the obstacles identified by Olson in his *Logic of Collective Action* (1971). A policy which will benefit everyone, and which confers no special benefits on the activists, does not provide the incentive for committed interest groups. Because anti-pollution policies, for example, may benefit everyone, there is no specific incentive for any one person to commit themselves to fight for tighter controls. This disincentive is reinforced by the tendency for producer interests to trump consumer ones. While campaigners against pollution are rarely its victims, those whose industry is under attack know that their jobs are at risk if the anti-pollution activists succeed. Furthermore, such activists are also vulnerable to the costs involved in gaining information. As we saw in the previous section, they cannot afford expertise and they may lack access to the relevant sources.

The weakness of the interest group appears yet more blatant when set in the context of the corporatist structure that characterises much high technology. Interest groups are liable to end up *serving* the needs of the corporate structure (Cawson, 1982; Dunleavy and O'Leary, 1987). Instead of providing a source of opposition, they become a means of managing dissent, of legitimising the political process, and of delivering policy decisions. Sandbach contends that popular participation, far from challenging the dominant interests, will actually reinforce them. His argument is that without 'stronger forms of popular control', participation only helps to consolidate the pre-determined policy decisions (Sandbach, 1980, p. 132). Deflecting such participation can often be crucial to the viability of the technology – when British Rail was contemplating the introduction of a new line, it realised that offering compensation to home owners along the track would add many millions of pounds to the cost of the project (*Observer*, 12 February 1989).

Alternative forms: between expertise and participation

Robert Dahl (1985) characterises the two approaches to decision-making we have examined so far as the choice between 'democracy' (public participation) and 'guardianship' (expertise). The guardianship option prescribes a non-participatory political system in which those with the relevant expertise (however this is defined) are given the responsibility for taking the key decisions. Technical knowledge, according to this argument, is disinterested, but complex. It requires expertise for its interpretation and understanding; it does not, however, include any ideological or value-laden element. Advocates of this view argue that there is a logic to modernisation; it is the embodiment of the idealised notion of science. Development and change are the fulfilment of the promise of technology. Traditional political choices – about the character of the 'good life' – have little part to play in this project, and it makes no sense to include them in the decision-making process. An early figure in this argument was Don Price (1965) who, whilst conceding some role could be played by politicians, insisted that their involvement be carefully demarcated and challenged by other claims, especially from those with knowledge and professional qualifications. By 'knowledge' he did not mean that of the representative about his/ her constituents, but of the expert about social and technical processes. For Winner, there is little left of democracy in Price's account: 'In the end, the system of government Price describes is not one that encourages or expects citizen participation, not one that relies upon any effective representative process, not one that includes much of a role for traditional politics at all' (Winner, 1977, p. 159).

The problems posed by a guardianship run by technical experts have already been mentioned. It is not clear that the idea of 'disinterested expertise' is actually coherent. But even if such expertise was available, there remains an important element of judgement in the weighting of the expertise and its application. As Dahl writes, 'virtually all important policy decisions require a judgement about the relative trade-offs between different values' (Dahl, 1985, p. 47). Or to put it another way, no decision is simply determined by the information available to the decision-maker and there is no point at which the 'information' determines the issue. The relevance of information has to be determined, as has the weight attached to it; and there are things, as we have seen, which are part of the decision but about which we cannot know the details. The question is, how should this judgement be organised and exercised? The assumption is that, given the need for judgement, the system should be modelled on democratic lines. The problem is whether it should be purely 'populist' or whether some important role should be found for expertise.

Populist democracy

The populist response may take a number of versions. Tony Benn (1982) and Leslie Sklair (1977) exemplify two forms of it. Their arguments are motivated by a desire to either control experts (Benn) or to deny them authority (Sklair). In both cases, popular participation, albeit in different guises, provides a bulwark against expertise.

Benn's approach is intended to strengthen the hand of the politician. He does this by devising 'Ten Questions for Scientists and Technologists' (pp. 96–7). They are:

1. How would the community benefit from the project?
2. What disadvantages are there to the proposal and who would experience them?
3. What skills are needed for the project?
4. Is there a cheaper less sophisticated way by which the objective could be achieved?
5. What new skills need to be created to operate the project?
6. What skills would be made obsolete?
7. Has the work been begun – or halted – in other parts of the world?
8. What disadvantages would accrue to whom if the project was not completed?
9. What other work needs to be initiated in order to cope with the initial project?
10. If a decision is made to proceed, how long will the option to stop remain?

The point of the interrogation is to establish the authority of the elected representative, and to introduce further dimensions to the decision (about how the technology will benefit the community, about the risks it bears and its impact on different social groups, about its effects on resources and employment, about its flexibility and controllability). The political representative seeks answers in order to ensure that the technology meets the criteria of democracy, which he/she embodies.

Sklair, by contrast, uses a more direct form of popular participation with the purpose of replacing experts. He attacks the assumption that 'the ordinary citizen' is incapable of making decisions about technology policy. He points to the array of complex decisions which people take daily. The reason people feel daunted by technology has nothing to do with their capacities, but rather with the way experts seek to preserve their professional power and status through the use of arcane language and demarcation rules. With the ground cleared this way, Sklair adds a further argument for popular participation. It is not necessary to know everything about a technology in order to make a good decision.

There are a number of weaknesses in the populist response. There might, in principle, be much to be said for Benn's ten questions, but the earlier discussion in this chapter suggests that they may not actually be answerable. Some element of judgement will be involved, and there then remains the problem of judging those judgements. Benn's approach, because it encourages immediate and definite answers, may in fact lead to an exaggerated certainty on the part of the experts. Or alternatively, the interrogation process will give undue weight to those experts who appear to deal in certainties (for example physicists), rather than those who work with probabilities (such as epidemiologists). Beyond this, Benn's strategy supposes the possibility of an overall perspective, a vantage point from which the technology can be judged. As we have seen, this too may not be possible.

If expertise cannot be corralled into serving democracy, there remains the option of substituting for it completely. If expertise is biased or incomplete, if an element of judgement is necessary, then why not give responsibility to those who will bear the consequences and costs of the technology – the public. Nelkin reports an instance of this kind of argument:

> After hearing 120 scientists argue over nuclear safety . . . the California State Legislature concluded that the issues were not, in the end, resolvable by expertise. 'The questions involved require value judgements and the voter is no less equipped to make such judgements than the most brilliant Nobel Laureate' (Nelkin, 1979, pp. 16–17).

There is an intuitive appeal to this populism, but it too has flaws.

Given the choice of two possible courses of action, the judgement of the person who has experienced both is worth more than that of the person who has experienced one or neither (Mill, 1972; Goodin, 1990). On the same basis, it might be contended that even if experts are wrong or disagree, they are wrong for better reasons than the layperson, who may make the right choice, but does so more by luck than by judgement. Experts also possess the advantages of having the background resources to learn from and thereby correct their mistakes.

Implicit in the populist argument is the idea that all complex decisions can be reduced to a core – political or moral – choice. This is a similar idea to the one which underpins the jury system. Although juries contain no legal experts, they exist to judge the culpability of a fellow human being. The point is, however, that the jury makes its decision within the terms set by the legal actors' interpretation of the law. There is no escaping the need for expertise in framing a problem, if not necessarily in answering it. Juries make their decisions on the basis of the evidence and the interpretations they are presented with (especially in complex cases, on tax questions, for instance); they do not decide how the case is to be conducted or the arguments marshalled. These key tasks are left to experts in these particular skills.

To acknowledge that non-experts can exercise judgement does not justify the claim that they should decide everything. As John Burnheim (1985, p. 49) contends, 'Even if we grant that the people can articulate what it finds objectionable in present practices and policies and the general direction in which they must be changed, it cannot articulate the concrete means by which changes are to be implemented'.

Or as John Dunn (1979, p. 19) argued more dogmatically, 'a political future which promised to dispense with expertise would be necessarily an idiot's promise or a promise made in the deepest bad faith'. The question then becomes how democracy and expertise might be combined.

Qualified democracy

The problem addressed by qualified democracy is that set by Michael Walzer (1985) – how to combine two claims to power: from 'those who know best how to use it' and from 'those who most immediately experience its effects'. Qualified democracy seeks to accommodate expertise by making it, in some sense, accountable and representative. Jennett, for example, recommends a consortium that is 'independent of established professional bodies and of government but able to talk to both. It should seek to distil data.... But it should also identify gaps in knowledge' (Jennett, 1986, pp. 272–3). A parallel line of argument is pursued by Dahl (1985). In wanting to retain the spirit of a populist democracy, but in recognising the impracticality of mass participation, he suggests the creation of a 'representative group of citizens', a 'minipopulus'. This small group stands in for its fellow citizens, doing their learning and thinking for them. The minipopulus would be advised by 'an advisory committee of scholars' – drawn from the National Academy of Sciences – who, says Dahl, are 'internally pluralistic and usually reflect the major intellectual currents and controversies of the field' and are 'probably about as independent of external controls as any organizations in our society'. With their advisers, the minipopulus would then make the judgements about the risks and uncertainties that arise in decisions about technology (pp. 84–8).

Qualified democracy works from the recognition that, on the one hand, decisions about technology are difficult and require considerable attention, and on the other, that they cannot be answered solely by recourse to technical expertise. Good decisions need the judgement and knowledge of ordinary citizens.

Whilst a qualified democracy acknowledges the limits of popular participation, it does not avoid all the obstacles posed by technology. In particular, the role ascribed to experts rests on a presumption of their independence. The character of modern technology tends to imbue expertise with a set of interests. Those with expertise in biotechnology also have an interest in the commitment of resources to their research; and those with the resources will have an interest in the research being applied. None of this need be taken as a form of corruption; it does, though, suggest that the expert cannot be presumed to be independent. The qualified democracy approach, like the populist variant, tends to take a rather superficial view of the actual business and problems of technology policy. Neither fully recognises the systemic character of much technology (Chapter 2) or the problems of assessment that we have raised earlier in this chapter. As a result the political proposals tend to miss the core problems.

CONCLUSION

We began this chapter by looking at the problems which affect decisions about technology, commencing with the general difficulties of biased choices, uncertainty and unknowability, before moving on to the more specific problems of risk and technology assessment and the competing political solutions to the problems of deciding about technology. The recognition that all such decisions involve judgements of some kind, stemming from uncertainty or value conflict, begged the question as to what principles should be applied. Both democracy and guardianship failed to cope fully with the problems that they were set.

A simple reform of the democratic process would not achieve an adequate system of control. It would fail to grasp the way in which technology shuts off certain options and sustains particular interests. Furthermore, it would have little to say about the problems of controlling technology. It is one thing to identify a 'good decision'; it is quite another to implement it. The problems of assessment, risk and uncertainty are too complex to be addressed in this way. While we have considered the question of what is required for the democratic control of technology, we are still left with the problems of judgement that are required by it.

Two possible lines of enquiry can be detected in what has been said so far. It may be that there is a technocratic solution to the problems of democracy, in which technology is itself used to meet the problems we have identified in choosing between technologies. The suggestion here is that human frailty can be replaced by a technical system. Rather than seeing technology as threatening democracy, it might be seen as furnishing the conditions *for* democracy. This solution is considered in Chapter 8. Alternatively, the answer may lie in political rather than technological reform.

Such thoughts inform the radical proposals of the Greens. Their argument is that the problems posed by technology for democracy and decision-making can only be solved by scrapping the technology, or by redesigning it to make it compatible with the principles of democracy. The Green response is the subject of the next chapter.

7 The Green Solution

INTRODUCTION

Most politicians and most political ideologies mention technology, but they rarely put it at the centre of their concerns. There may be reference to the need for technological development or the inadequacies of a particular example of technology, but such comments are usually incidental to the main thrust of the political philosophy. The exception to this general rule is, of course, the Green movement. The Greens' argument displays a deep mistrust of existing forms of technology, and our reliance upon them. At its starkest, the Green analysis doubts whether democracy and current technology can co-exist. To salvage democracy, much technology needs to be renounced or radically reformed. This chapter examines the Greens' attack on technology and, in particular, their solution to the problem of linking technology and democracy.

One of the most striking political phenomena of recent years has been the rise of the Green movement (Lowe and Goyder, 1983; Ward, 1990). The last two decades have seen a massive explosion in membership of Green-related groups. In Britain, this can be seen in the rise of membership of organisations like the Council for the Protection of Rural England (CPRE), Friends of the Earth (FoE), the Ramblers Association (RA) and the World Wide Fund for Nature (WWFN) (see Table 7.1).

Table 7.1   Membership of British Green-related groups

	Membership (000s)		
Group	*1971*	*1981*	*1989*
CPRE	21	29	40
FoE	1	18	140
RA	22	37	75
WWFN	12	60	200

Source: Social Trends 21, 1991.

Their dramatic increase in membership tells only part of the Greens' success story. In other countries the Greens' popularity has been translated into political representation. The first Green politician was elected to a national assembly in Switzerland in 1979. In 1983, the German Bundestag contained 28 Green members. By 1984, Green Parties were collecting three per cent of the vote in the elections for the European parliament. Five years later, the Greens won 15 per cent of the European parliament vote in Britain alone. There have, of course, been fluctuations in the Greens' political success (the German Greens [*Die Grunen*] performed very badly in the 1990 German national election), but their presence on the political scene seems to have been established, if not numerically at least in terms of the political agenda. This can be seen in the way in which environmental issues have taken on political salience. All political parties feel compelled to make some reference to the environment, just as every manufacturer now has to persuade consumers that their goods are environmentally friendly.

DIFFERENT SHADES OF GREEN

To observe these trends tells us little about the political values and ideas that underpin them. There is no simple relationship between the rise of the movement and the values it espouses. While there maybe, as Inglehart (1977) suggests, a shift towards post-materialist values, and while the Greens may represent those values, it does not follow automatically that the one explains the other. Many contingent and structural factors play a part in the passage from ideas to action. Our concern here is primarily with the ideas.

In analysing the Green argument, we have to be aware that there is no single Green movement, and there is, therefore, no single 'Green' analysis of technology. Not only are there differences between Green political parties, there are also divides within those parties. There are splits within *Die Grunen*, for example, between those who are prepared to compromise with the existing power structure (the 'Realos') and those who reject it totally (the 'Fundis'). Their differences can be explained by their political interests (the Fundis dominate the Party executive, while the Realos tend to predominate among the elected representatives), but this does not alter the fact that they adopt divergent interpretations of pollution and the solutions appropriate to it. For the Realos, there is much to be gained by reforming the existing system; for the Fundis, the only answer lies in a radical transformation of the whole of society. Jutta Ditfurth, a proponent of the latter view argues, 'Reformism, in its Green–Social democratic form, starts out from the assumption that there will be no radical break in our society. . . . For any serious analysis must take into account the fact that every variant of reformism has proven itself historically bankrupt' (quoted in Parkin, 1989, p. 124). But the debate between Greens need not necessarily conform to the familiar divide between reformists and revolutionaries, if only because Greens, in all their incarnations, reject the traditional model of the political party that underpins that debate. They argue that a political party is not simply a machine for winning votes, nor for producing policies or blueprints; instead it is representative of an *approach* to policy and politics, the embodiment of a set of *values*.

Being 'Green' does not automatically place you at any particular point on the political spectrum. Consider these two quotations:

> The war which man is now waging against nature is not only foolhardy, it is downright dangerous. We can't go on polluting the atmosphere, we can't go on poisoning the sea, we can't go on slaughtering our wildlife for ever.

> [I]f you look at the net demand we are making on the planet and project the needs of people over the course of time against what the planet can meet, it's evident that at some stage in the future we are going to come to the end of the present consumption of finite resources. Either we cause the end ourselves, or it happens through some means which will be extremely unpleasant to live through.

Although both quotations express superficially similar sentiments, they emerge from quite different sources. The first appeared in *Nationalism Today*, published by the British right-wing, racist party, the National Front (Benton and Edwards, 1984). The second appeared in *Marxism Today*, the magazine of the British Communist Party, and comes from an interview with Jonathon Porritt, a leading member of the Green movement in Britain (Porritt, 1984b, p. 25). The fact that apparently similar arguments can emerge from diametrically opposed political positions suggests that we should be wary of glib generalisations about 'the Greens'. For the extreme right, environmental concern represents a desire to return to a feudal order. For the left, protecting the environment represents a means of ensuring the conditions for an egalitarian, participatory society. In analysing Green arguments we need to be aware that entirely opposite political assumptions, values and strategies may be grouped under the same heading. This applies not only to groups at either end of the political spectrum but also to people in the same party – almost every Green party echoes the debate between Fundis and Realos.

Some critics of Green arguments would contend that, whatever their apparent differences, Greens on both the left and the right are reactionaries driven by a romantic vision to seek undesirable ends by impractical means. While criticisms of this kind may indeed focus on important weaknesses in both sides, it is of little help (except as a polemical device) to link the two. Their political differences significantly alter the case being made. Even if both sides look to the past for their alternative to today's industrialised world, and even if they romanticise that past, they provide quite opposite pictures of how things ought to be and how this can be achieved. As Rosalind Williams observes 'mourning for the forgotten or ruined landscape is not necessarily a sentimental and outdated emotion. It can also express an ideological response to tyranny' (Williams, 1990a, p. 147).

Given the focus here on democracy (and not feudalism), I shall not dwell on the right's version of the Green vision. Instead I shall concentrate on the ideas of Jonathon Porritt and others, who sit to the left of the conventional political divide. It is, after all, their form of the Green argument that is most widely recognised, even if it is rarely understood.

The kind of Green ideology we are considering entails more than a commitment to recycling waste and to purchasing environmentally-friendly products. Porritt and others see themselves as 'Deep Greens', or in the terminology devised by O'Riordan (1977, p. 4), they are 'ecocentric' as opposed to 'technocentric'. Ecocentrism argues that 'social relations cannot be disconnected from man-environment relations'. Technocentrism, by contrast, takes 'a man-centred view of the earth coupled with a managerial approach to resource development and environmental protection'. Where technocentrics work within the existing political structures and seek to correct the results of the current system, the ecocentric seeks a change in that system which would lead to 'a redistribution of power' and 'a federal political economy'. Though this distinction does seem to fit the reformist–revolutionary divide, it incorporates a far wider range of issues than is normally associated with that debate. Andrew Dobson, who draws a similar distinction to O'Riordan but labels it differently, also points to the sweeping character of changes anticipated by the Deep Greens' position. Where O'Riordan talks of ecocentrism, Dobson prefers 'ecologism':

> Ecologism seeks radically to call into question a whole series of political, economic and social practices. . . . While most post-industrial futures revolve around high-growth, high-technology, expanding services, greater leisure, and satisfaction conceived in material terms, ecologism's post-industrial society questions growth and technology (Dobson, 1990, p. 205).

It is this attitude to technology that is our focus here.

THE GREENING OF TECHNOLOGY

Technology lies at the heart of Green ideology. In his survey of Green political thought, Dobson observes that, while Greens are not united in their view of technology, 'What can be said, it seems to me, is that whole hearted acceptance of any technology disqualifies one from membership of the Green canon' (Dobson, 1990, p. 98). Technology stands as the key element in dividing the 'real world' of nature from the 'surrogate world' of artifice. And the main aim of Green political activity is to return society to the 'real world', to recover a natural, organic order.

Such arguments can be seen in Porritt's book, *Seeing Green* (1984a). Porritt has three main purposes. First, he wants to make the case for environmental concern – why we should care. Secondly, he seeks to explain how the existing regime has promoted the environmental destruction of the planet. Finally, he makes the case for a Green movement and a Green society.

The pollution of the planet and the destruction of its resources are more than acts of vandalism; they thwart the development and realisation of a natural balance between the planet and its inhabitants. Porritt concludes that 'we must learn to blend our concern for people with our respect for the Earth through the post-industrial politics of peace, liberation and ecology: the politics of life' (ibid., p. 235). A similar theme is developed in the British Green Party's 1987 election manifesto: 'Green politics is about building a new way of life, one that is based on respect for our planet and humility about our role in it. We need to stop building on the quicksand of materialism, patriarchy, competition and aggression' (Green Party, 1987, p. 1).

The cause of environmental pollution, according to Porritt, is neither socialism nor capitalism; it is industrialism. He defines this as an overriding commitment to material growth which is expressed through an economy 'that attempts to meet people's needs by ever-increasing production and consumption' (Porritt, 1984b, p. 25). Industrialism, though, is less an economic structure and more a state of mind. In this sense, Porritt is offering a cultural explanation of the decline of the environment, and as a result he looks to cultural change to halt it. The book's opening epigram quotes Rudolph Bahro: 'When the forms of an old culture are dying, the new culture is created by a few people who are not afraid to be insecure'. Change is to come through the replacement of old attitudes and ideas. Porritt rejects any crude historical materialism which sees environmental destruction as a consequence of 'capitalism'. State socialist societies are no better. Porritt is also sceptical of change through conventional political parties. He favours movements which embody a new order and a new political culture.

Industrialism, for Porritt, is an ideology which envelops and incorporates 'modern science and technology', which have been shaped to serve industrialism. Science 'is simply not geared up to cope with the priority problems of humanity', while existing technology is 'inappropriate' and 'destructive'. Political structures also serve to maintain the grip of industrialism. Dominant industrialist interests are represented through a technocracy which organises the political agenda to make certain technological developments 'necessary', most obviously in the creation of nuclear arsenals (Porritt, 1984a, pp. 50, 59).

The industrialist ideology, while powerful, is neither inevitable nor invulnerable. There is an alternative: 'It is a matter of choice whether technology works to the benefit of people or perpetuates certain problems, whether it provides greater equity, and freedom of choice or merely intensifies the worst aspects of our industrial society' (ibid., p. 130). The choice is exercised in order to reassert human values. It is not, however, just a matter of finding new uses for existing technologies, any more than it is a matter of redirecting the existing political structure. The key is to devise technology that is appropriate. There are two elements to the idea of 'appropriate'. At one level, it means designing technologies that are controllable, that are 'democratic'. This recalls Lewis Mumford's distinction between authoritarian and democratic technics:

> What I would call democratic technics is the small-scale method of production, resting mainly on human skill and animal energy but always, even when employing machines, remaining under the active direction of the craftsman or the farmer (Mumford, 1972, p. 52).

Authoritarian technics, by contrast, are huge, complex and centralised, defying human control. The second dimension to 'appropriateness' refers to results of technological development, the need for socially-useful products which fit with the general democratic structure. As the 1987 British Green manifesto puts it: 'We will use scientific research funds to explore technologies which are most likely to serve a decentralised, sustainable society.' (Green Party, 1987, pp. 4–5). The implications of such a view are spelled out in the report, *Our Common Future*, by the World Commission on Environment and Development (1987, p. 60):

> In all countries, the processes of generating alternative technologies, upgrading traditional ones, and selecting and adapting imported technologies should be informed by environmental resource concerns. Most technological research by commercial organisations is devoted to products and process innovation that have market value. Technologies are needed that produce 'social goods', such as improved air quality or increased product life, or that resolve problems normally outside the cost calculus of individual enterprises, such as the external costs of pollution or waste disposal.

The consequence of such recommendations is that public policy, if not the political system as a whole, should create incentives to behave in environmentally responsible ways.

Given that there are alternative orders available, the task of the Green movement is to create a situation where these other worlds can be selected. The solution lies with democracy; that is 'restoring power to the community'. 'Having control over technology', said the British Green Party in 1987, 'means deciding when, where and how to use it, whether to develop a particular technique, whether the side effects justify its use, and if necessary when to stop. That power must be in the hands of the millions of people who will be affected by the choice' (Green Party, 1987, p. 4). For Porritt, democracy means a participatory communalism. Representative democracy has failed the environment by allowing the separation of the people from their representatives, and those representatives have lost touch with the real interests of their constituents: 'the *representative* element of the system has insidiously undermined the element of *participation*' (Porritt, 1984a, p. 166, his emphasis). The structure and role of politics serves to reinforce the commitment to industrialism. Neither constitutional nor institutional reform – whether it is proportional representation or the reorganisation of the administrative bureaucracy – will bring the necessary control. Any attempt to create change from within is doomed. The only alternative is a gradual change of ideas and values, sustained by small-scale experiments at the local level. Decisions have to be taken locally and have to be mirrored by a decentralised economy organised around local savings banks and regional enterprise boards (ibid., p. 134). Porritt is not, he claims, advocating parochialism; he is committed to the familiar Green slogan: 'Act locally, think globally'.

These proposals would have dramatic consequences and would require much more than a redistribution of power. Consider, for example, Andre Gorz's solution to the car and its associated problems of pollution, impracticality and congestion:

> Unlike the vacuum cleaner, the radio, or the bicycle, which retain their use value when everyone has one, the car, like the villa by the sea, is only desirable and useful insofar as the masses don't have one. That is how in both conception and original purpose the car is a luxury good (Gorz, 1980, p. 69).

Gorz concludes that access to the car cannot be 'democratised' – 'If everyone can have luxury, no one gets any advantages from it'. But if the car cannot be democratised, what can be done? It is not, says Gorz, simply a matter of choosing to do without the car. There is no choice: 'You aren't free to have a car or not because the suburban world is designed to be a function of the car – and, more and more, so is the city world' (ibid., p. 75). Equally, technical fixes, in the form of bicycles, buses and the like, are ruled out for the same reason. The design of the city makes them unviable. Instead, Gorz proposes a radical solution in which people give up transportation altogether.

The acceptability of this answer depends on totally redesigning the community so that people no longer have the *desire* to travel (ibid., p. 76). What is being claimed here is that technologies which are environmentally dangerous or damaging, or are beyond effective reform or control, must be eliminated. The process of elimination does not mean just consigning the technology to the scrap-heap. It means reconstituting the need that first brought it into existence. Such thoughts lie behind Porritt's call for an examination of the difference between 'genuine needs' and 'artificial needs'. The distinction, he says, rests upon 'an unequivocal value judgement': 'Is our well-being genuinely enhanced by electric toothbrushes or umpteen varieties of catfood? Indeed, is it genuinely enhanced by having "unlimited" freedom of choice?' (1984a, p. 197).

In summary, the Greens' argument with modern technology derives from the movement's general political approach. Technology is the proximate cause of environmental desecration, but its role is shaped by the 'desires' and 'needs' that fuel its creation. These false ambitions are sustained through the political and economic structure of industrialism. The Greens' aim is to save the environment by overturning the values and judgements which allow modern technology to flourish, and to replace them with a moral order based on genuine needs and sustained by a locally-organised, participatory political economy.

CRITIQUES OF THE GREEN ARGUMENT

Just as there is no single 'Green ideology', so there is no one counter-argument. The critiques come from a variety of political positions with different emphases. We focus first on the structural critique which argues that the Greens fail to acknowledge the economic structure that shapes the possibility for change.

Structural critique

Francis Sandbach (1980) argues against the idea that culture holds the key to change, which strategy the Greens adopted as the logical consequence of having attributed the cause of environmental decay to 'industrialism'. They see industrialism more as a state of mind than a definite political structure, and they contend that this explains why different structures produce essentially the same problems. Sandbach doubts that changes in attitude and values can lead to a change in the technology adopted by a society. For him, such thinking is Utopian and fails to recognise that existing forms of technology and the attitudes which surround them are generated by the economic structure. Changing technology and attitudes depends, therefore, on changing the industrial determinants of both. 'To establish alternative technologies', Sandbach writes (1980, p. 182), 'political campaigns must become part of the main labour movement's struggle to reduce the influence of the capital owning class'. Sandbach's criticism is not directed at the values to which the Greens subscribe. Rather, his claim is that, whatever the values, change is determined by access to, and ownership of, the means of production. It is, however, not clear how vulnerable the Greens are to this particular attack.

While it may be true that present technologies sustain capitalist interests, it does not follow that this *explains* their existence. Were this so, we would be able to see a clear connection between any given technology and the interest it served. But the same technology may be found in quite different contexts, and have quite different effects. Mass communications technology, for instance, can be a source of manipulation or liberation, depending on who employs it and how it is organised.

Even if a technology is the creation of a particular set of class interests, it does not follow that the only way to change that technology is to organise a countervailing set of interests. The emergence of new information can itself generate a change in the perception of the costs and benefits of that technology. The Greenhouse effect, for example, cannot be attributed to any particular set of interests, and yet it has had some, albeit limited, effect upon the practices of both government and industry. This is not to deny that the creation and dissemination of information is itself subject to political controls, only that its capacity to change behaviour is not conditional upon a simple class conflict model.

Equally, we need to distinguish between the intentions and the effects of a technology. While it may be that a particular set of interests caused the creation of a technology, it does not necessarily follow that its subsequent use will reinforce those interests. This may not be true for all technologies, but it is for some. An instance of this is the cassette recorder, which began as another means for marketing music but became a device by which the power and resources of record companies were threatened through the rise of home copying and independent recording. In short, the structural critique of the Green argument tends to overlook the ambiguities in the development of technology, the divergence between intention and effect, and the impact of contingent changes in its environment. All of these suggest that any attempt to give an unqualified purpose or form to technology may obscure more than it clarifies.

Class critique

A variant on the structural critique of Green politics is the view that the movement represents a set of material interests rather than disinterested concern for the planet. This line of criticism begins with the observation that the movement is predominantly middle class (see Allison, 1975; Rudig and Lowe, 1986). The critical inference drawn from this is that support by the middle class for the Greens is given because of the *exclusive* benefit which is afforded that class by campaigning under the Green umbrella. In short, the Green movement represents no more than an interest group for those whose privileges are threatened by the spread of motorways and tourism. Behind the universalist rhetoric of Green concern, with its talk of 'only one world' and 'our common future', there lies, it is suggested, a very partial perspective. The ideology of a common plight caused by a common problem disguises a reality marked by competing sectional interests. To attribute concern for an environmental problem to an undifferentiated mass called 'humanity' overlooks the differential effect of the problem and imposes a set of priorities which may be very specific. Even if pollution is encountered universally, it does not follow that either its cause or its effect are common. The 'concern' of the middle class may, therefore, marginalise or deny the experience of other groups, shifting the political agenda to serve one set of interests over another.

A similar critique is expressed by Hans Magnus Enzensburger, who finds in talk of 'spaceship earth' an ideology which serves to advance the interests of those who presently dominate society. 'Global concern' acts as cloak which hides conflicts of interests, just as does the use of terms like 'the national interest'. 'One of the oldest ways of giving legitimacy to class domination and exploitation', writes Enzensburger, 'is resurrected in the new garb of ecology'. It is not Enzensburger's contention that the Greens are wrong about the problems of scarcity and pollution. His point is that their analysis of the cause of these problems and their solutions to them are both partial. In particular, he speaks of the missionary zeal with which they paint a picture of disaster and decline, and then produce, as if by divine revelation, a solution which merely involves a change of heart by all rational, right-minded people (Enzensberger, 1976, pp. 253–95).

The class critique, like the structural one, develops an important line of thought. Certainly, expressions of concern for the environment cannot be taken at face value. The assumptions and values behind the concern must be unmasked. On the other hand, no simple connection can be made between beliefs and interests. People can genuinely argue for causes that run counter to their interests, just as commitment to a cause that is in one's interest does not invalidate the changes being called for. Assessing the Green argument, therefore, requires a closer examination of its politics, of the values it espouses and the solutions it proposes.

Political critique

The previous criticisms raise the question of the form and the direction of the changes which the Greens seek. A major tension at the heart of the Greens' vision of democracy is contained in the idea of thinking globally and acting locally. While the Greens' 1987 manifesto emphasised the party's commitment to local democracy, it acknowledged that local democracy depends upon local power. For this, it would be necessary for transnational corporations to be controlled by 'A strict Code of Conduct'. Such a code would have

to be operated in a world in which local government was responsible for the powers given to central government. Two problems
arise from this. Does a federal system provide the kind of central
authority which would be able to challenge the power of transnational companies? And would local democracies be able to meet the
demands of its citizens?

These problems become more complex when set against the
Greens' 'global' vision. The problems which the Greens aim to
tackle are increasingly portrayed as transcending national, let alone
local, boundaries. O'Riordan, for example, writes:

> Global warming should be neither a party-political nor an
> electoral issue. Its causes are not politically based: all countries,
> with every variation of political ideology, contribute to some
> extent. The management of global warming requires the rise of
> the global politician, buttressed by a global citizenry, whose
> vision extends for decades, rather than the next election
> (O'Riordan, 1990, p. 12).

Whether or not this plea is either feasible or desirable, it is clearly in
conflict with the demand for local democracy. But there are further
grounds for doubting the Greens' ambitions for decentralisation,
not so much because of its undesirability as its impracticality. Not
only may it be very difficult to decentralise certain technologies; to
do so may also be counter to Green political principles. Forms of
mass communication, for example, have to be organised across
communities and require some measure of coordination and
regulation. Without mass communication there can be no sense of
the globe about which the Greens are concerned. The problem does
not just apply to intrinsic features of the technology, but to the
social structures within which they are embedded. Hollick writes:

> Probably the strongest argument against decentralisation is its
> sheer impossibility in the foreseeable future. The majority of
> people in developed countries already live in cities, the develop-
> ing nations are urbanizing rapidly, and there seems little prospect
> that this trend can be reversed at least until population growth is
> controlled. Once established, cities represent a capital stock of
> housing, industry, commercial premises and services which turns
> over relatively slowly and cannot simply be abandoned or

replaced. They also contain much of the cultural heritage of any civilisation that is essential to social identity and cohesion (Hollick, 1987, p. 293).

The inescapability of dependence upon technology may be precisely the problem upon which the Greens focus, but there is no particular virtue in identifying problems that have no solution.

There are those who would argue that there is a way out, that alternative communities can be evolved and that a self-sufficient technology can be created. But even if such escapes were available to all, the evidence of their success is mixed. One experimenter in technological self-sufficiency, Robin Clarke, reflected ruefully on his experiences:

> Suffice it to say that I and my family left 18 months [after arriving], and another five a few months after that. Nearly all of us left for the same reason: the struggle to do things we wanted against a background of mounting inertia and community dissent proved too great. Just over three years later the community had been officially disbanded, and the farm sold (Clarke, 1976, pp. 13–14).

Interestingly, the lesson Clarke draws from his life in a self-sufficient community is not that such an existence is physically impractical; the technology can be made to work and a tolerable standard of living can be enjoyed. The problem is in the politics, in the business of getting along (ibid., p. 284). Dunn echoes this point in the wider context of introducing appropriate technology into less-developed countries. 'No action is politically neutral', he observes, 'not even installing a pump, since the effect may be either to strengthen or weaken the existing social system' (Dunn, 1978, p. 39). The Green case takes too little account of the political problems posed both by existing relationships and by planned future ones.

Such obstacles are all the greater when considered in conjunction with Hirsch's identification of the social limits to growth. Growth, he argues, is measured by the increase in access to desired goods. The difficulty is that some goods cannot be increased in this way; in particular, positional goods. A positional good is one which depends for its value on the exclusion of others from enjoyment of it – for instance, peace and quiet or an unspoiled view. Any attempt to increase access to the positional good is liable to destroy it. Hirsch takes the example of the city suburb: 'The attractions of suburban living provide incentives to both city dwellers and country dwellers to move to the suburbs. This process of movement will in turn change the characteristics of suburban life, at first to its net benefit but after some point to its detriment' (Hirsch, 1977, p. 37). Access to goods such as those offered by the suburbs must be managed, whoever is to enjoy them. The pleasure they give depends on the limits placed on access to them. Such management cannot be achieved through competition by local communities because they would simply reproduce the competition which threatened to destroy the goods in the first place. The net effect would be the destruction of the good itself. Hirsch's lesson for the Greens is that there are goods, often ones which are environmental, which cannot be distributed universally. Furthermore, protection of such goods requires externally imposed forms of rationing. In short, concern for the environment and commitment to democracy may be in conflict with each other.

It is, of course, part of the Greens' case that the political system encourages popular demand for economic growth. This growth brings with it unwelcome consequences for the quality of life and there are no incentives to counter these tendencies. Politicians get elected on the basis of their promise to give us more, not less, of whatever it is that we want. In the process we all internalise a desire for 'more', a desire which the Greens want us to eliminate. A new political order will have to create a new set of more reasonable demands. But is this argument coherent?

It depends on the assumption that democratic systems do, in fact, generate an environmentally destructive form of collective behaviour in their electorates. Lauber (1977–8) challenges this assumption, arguing that the 'demand' for growth is the creation of certain elite interests, and cannot be attributed to a general populist cultural ethos. While growth advances the power of economic elites, the self-interest of individual citizens may actually work against growth. The Greens' own political perspective, with its focus on the individual and their attitudes, leaves little space for structural processes and the possibility that individual views and values may be the only coherent response to these external forces. If the Greens misplace the responsibility for the demand for growth, then their solution (cultural change plus democratic restructuring) may be unnecessary or inappropriate. This reinforces the earlier line of political criticism which focused on the tension between global problems and local solutions.

Moral critique

But even if the political reforms proposed by the Greens are viable, there remains the question as to whether the alternative world they offer is itself capable of generating the change of heart they call for. The Greens seem to be proposing one of two possible deals to us – they both concern the moral worth of the future, Green world. Either they are suggesting that their society would be qualitatively different from our existing one by virtue of our changing sense of who we are and the practices we adopt – put simply, we will cease to be acquisitive, competitive and materialist. Or we remain roughly the same people, but we come to see the new Green world as better fitting our aspirations (aspirations that were frustrated by the old order). A comparison between the Green promise and the industrial present shows that this would necessitate some kind of trade-off – possibly that the future benefits of political participation and control will outweigh the loss both of present economies of scale and of the range of available products and services.

Each of these two deals might be questioned. If we were to be totally changed by the adoption of the Green alternative, then we could no longer reflect rationally upon the pre-Green world that we inhabited – there would be no common standard of comparison and therefore no way of knowing whether we were better off. The other deal at least allows for comparison. What is not clear, though, is whether we would in fact choose the Green alternative, whether we would feel ourselves better off in the communal existence offered by the Greens. While there is considerable evidence that the rise of the Green movement has engendered changes in awareness and habits, this does not necessarily indicate a willingness to adopt a radically different form of life. Is it reasonable to expect, as Gorz seemed to, that people will lose the 'desire' to travel, if this is the condition required for the elimination of the car? In default of any such change of heart the Greens might be faced with having to persuade people that morally they *ought* to give up their car, whatever it is that they might otherwise wish.

This sense of moral obligation is best expressed in the Greens' appeal to the advantages of establishing a 'natural' order. It is captured in words like 'wholeness', 'organic', 'web', or in statements like this: 'The Earth has been served by the wisdom of ecology for millions of years. We can use that wisdom to make us whole again. With it we can restore the balance between the logical and the natural, between ourselves and other people, between all humanity and the planet Earth' (Green Party, 1987, p. 1). There are two elements which might be used to give authority to this moral claim. The first lies with the contrast they make between the 'natural' and the 'artificial'. The second depends upon the idea that the 'natural' is superior. The distinction, and the judgement attached to it, lies in their suspicion of technology. It also has a general plausibility in that it fits with many commonly shared intuitions. Countless advertisements play upon the idea that what is natural is good – the 'natural goodness of wholewheat', and so on. Many aesthetic judgements are informed by the thought that 'authenticity' (and its correlate 'integrity') are essential to the business of self-expression. It is no coincidence that critics of popular culture choose the label 'plastic' to condemn it. But even if the Greens have tapped into a widely shared 'common sense', it does not follow that they are right in their judgements and that there is any clear correlation between the natural and the good.

Keith Thomas' historical study *Man and the Natural World* reminds us that the idea of 'nature' is changing constantly, and that these shifts are related to a view of the world. 'By the late eighteenth century the appreciation of nature, and particularly wild nature', writes Thomas (1984, p. 260), 'had been converted into a sort of religious act. Nature was not only beautiful; it was morally healing'. In other words, the understanding of 'nature' was selected rather than given. We need to think only of Thomas Hobbes' 'state of nature' to see another, albeit quite different, use of nature for political ends. For Hobbes nature is a threatening, unruly thing. Whatever perspective is adopted, 'nature' is used ideologically to sustain or legitimate a state of affairs. Nature does not simply speak to us. It is interpreted through political ideals. As Winner comments (1986, p. 137): 'Nature will justify anything'.

The Green moral argument also rests upon a particular categorisation of the 'natural' world. It sees the world in a particular way, as a set of life forms (animals, plants, humans), and then seeks to establish that each form is entitled to equal status. Richard Sylvan defines this approach as an 'elaboration of the position that natural things other than humans have value in themselves, value sometimes perhaps exceeding that of or had by humans'. Deep ecology can then be seen, says Sylvan, to challenge the 'sole value assumption' of Western philosophy (which underpins social and economic arrangements), that 'humans are the only things of irreducible (or intrinsic) value in the universe' (Sylvan, 1984, pp. 2–5). This general Green assumption is, however, difficult to apply; it begs a further series of awkward questions which threaten to force the Greens into an indefensible position. Sylvan asks why should our planet be singled out? Are animals of the same status as rocks? William Grey is similarly critical of Green philosophy in its deep ecology mode. Among his criticisms is the argument that 'not all primitive resource-use is wise, and not all technology is destructive'. To establish what practices are sound, we should not reject science but employ it: 'Nature may indeed know best, but how, except through systematic empirical inquiry, can we determine what it is that nature tells us?' (Grey, 1986, pp. 211–16).

If we see the Greens' reading of technology as having been created from within a particular interpretation of 'nature', so that both nature and technology are understood ideologically, then the idea that nature is 'natural' and technology 'artificial' makes little sense. They are both artifical. So that whatever criticisms can be made of technology, the claim that it is at odds with a natural order is incoherent. The criteria for judging its appropriateness has to be grounded in political dialogue rather than in assertions about nature.

CONCLUSION

The Greens view technology warily, if not with open hostility. It represents a source of problems, particularly when it is allied to a commitment to industrialism. If people are weaned off the culture of industrialism, if they adopt a decentralised political structure, if they are encouraged to develop appropriate technology, they will recover a harmonious, natural balance with the planet. These claims provoke, as we have seen, a number of criticisms. These argue that the Greens underestimate their opponents (the structural critique), or that they serve malign interests (the class critique), or that their solution is riven by irreconcilable tensions (the political critique), or that their general arguments are built upon shaky foundations (the moral critique). While each of the critiques may be rejected, they cannot be ignored when trying to establish a coherent politics of technology.

Even if we were to agree that the Greens' vision is a desirable one, we still have to ask whether it is coherent or possible. The claim that Green ideas have percolated into daily life and into mainstream political parties is undoubtedly true. The presence of Green issues on the national political agenda is confirmation of this, as are the changes being made by commercial companies to their product range in order to accommodate a new consumer awareness of pollution and other hazards. Such developments help sustain the idea that there has been a general shift towards the ideas and values of the Greens. But in themselves these changes do not coalesce into a coherent political movement. The link between theory and practice seems especially fragile. It is unclear whether local changes can indeed transform global political and economic structures.

But even if the Greens have access to the source of social change, we have to judge the subsequent political system. It is important to ask what political forms and relations will evolve in the post-industrial order. Will a nation of self-sacrificing altruists be able to make tough local choices which may have adverse effects on them but promise some general planetary improvement? Opinion poll data regularly reveals that people believe in the need for general national change, while at the same time showing no willingness to adapt their private lives. They extol public transport and remain wedded to the car. The Greens seem to lack any mechanism for overcoming these awkward facts, facts which, as the Greens themselves recognise, are the logical expression of the 'problem' of industrialism. Dobson comments (1989, p. 46), 'the politics of ecology does not follow the same ground rules as its philosophy'.

The problems of choice and political structure which the Greens confront are not exclusive to them. They infect almost all accounts of democracy. It is perhaps appropriate that, in the next chapter, we consider the argument of those who see the problems of democracy being solved by the very thing that the Greens treat with such suspicion – technology.

8 The Technical Fix

INTRODUCTION

During the 1991 Gulf War, the Iraqis began discharging Kuwaiti oil into the sea, causing massive pollution of the water and coastline. The US airforce solved the problem by bombing the pipe outlets. This is one example of a 'technical fix'. Here is another. People plagued by obscene or unsolicited phone calls have devised a useful technique for countering them: to blow a loud whistle down the mouthpiece. This chapter is about technical fixes, about the use of technology to solve political problems.

The Greens tend to assume that technology poses problems rather than solving them. For them, the technical fix is almost always an inadequate response and may even be a disastrous one. For other commentators, though, the application of technology provides a genuine answer. *The Economist* (16 February 1991) remarked:

> It is one of the curious things about the recent fascination with global warming that so little interest or faith has been shown in technology's ability to solve the problem.... For example, some scientists recently calculated that the growth of plankton in the Antarctic ocean was limited by the supply of iron. Since plankton gobble carbon, why not fertilise them with cheap iron filings and thus reduce global warming? Greenpeace was horrified. A technical fix! What might it do to the sensitive ecology of the Antarctic? It might upset the natural balance. Yet, who knows, it might be a lot cheaper and more politically manageable than reducing fossil-fuel use.

The Greens place technology and democracy in conflict. The survival of one jeopardises the other. Proponents of the technical fix foresee no such danger, and even claim that democracy *requires* the application of technology.

Certainly, if, as the previous chapter suggested, 'nature' is a social construct, we can no longer make a clean break between what is natural and what is artificial; we cannot use the artificiality of technology as an argument against its deployment. Put another way, technology can be seen as 'natural'; it can be represented as the realisation of the human capacity for self-expression and development. Or technology can be equated with freedom because it 'tames' nature's powers. In adopting such arguments, technology and democracy cease to be alienated from each other. Technology, like democracy, is part of the attempt by human beings to control their world. Instead of technology posing problems for the democratic order, technology provides for the social control which is implicit in 'popular rule'; it takes control away from nature and gives it to the people. That, at least, is the idea.

This chapter asks whether it makes sense to see technology like this, as the friend not the enemy of democracy. We need to begin with the core idea: the 'technical fix'. Pacey (1983, p. 7) defines the technical fix as 'an attempt to solve a problem by means of technique alone'. O'Riordan (1977) brackets it with the idea of 'technocentrism', contrasting it with the environmental movement's ecocentrism. Technocentrism expresses the notion that technological development is essentially an exercise in problem-solving. As such, it chimes with much that passes as 'common sense'.

Technology enables people to do what they want. The expansion of these opportunities represents 'progress'. The measure of civilisation is the access people enjoy to certain technologies – cars, televisions, refrigerators and so on. Equally, it is commonly assumed that, with suitable technology, any problem can be solved. The coalition forces in the 1991 war against Iraq were thought to be guaranteed victory by the superiority of their military hardware. Another example of the technical fix is contained in this newspaper report:

> Three cabinet ministers have computers to thank this month as radical policies nearly 10 years in the making fall into place. . . . [A]mbitious computerisation projects . . . have delivered the flexibility they need to revolutionise our tax, welfare and health services (*Independent*, 28 March 1988).

The implication of this report is that the computers have made political change possible.

These examples of technical fixes are part of a general view that technology is the key to the realisation of human goals. It is a view that is not confined to any one political tradition; it is evident in both Marxism and liberalism. Marx's account of the history of human development incorporates the notion that technology is a progressive force, causing improvements in ability to control the natural environment. In the *Communist Manifesto* (1967, p. 84), Marx and Engels proclaim: 'The bourgeoisie, by the rapid improvement of all instruments of production, by the immensely facilitated means of communication, draws all, even the most barbarian, nations into civilisation'. And they go on to explain how

> The bourgeoisie, during its rule of scarce one hundred years, has created more massive and more colossal productive forces than have all preceding generations together. Subjection of Nature's forces to man, machinery, application of chemistry to industry and agriculture, steam navigation, railways, electric telegraphs, clearing of whole continents for cultivation, canalization of rivers, whole populations conjured out of the ground – what earlier century had even a presentiment that such productive forces slumbered in the lap of social labour? (ibid., p. 85)

Although Marx acknowledges the cost of the changes engendered by the development of technology, he thought that the new techniques were essential to the ultimate emancipation of the working class. The machine represented the ability to tame nature and to establish the conditions for freedom.

This vision has remained remarkably constant, if not universal. It can be read into the 'end of ideology' literature that emerged in the aftermath of the Second World War. Then, the progressive ideas of science were thought applicable to all aspects of society, including politics. Writing in 1948, the scientist C.H. Waddington reflected on how the scientific approach could solve society's problems:

> Its first step would be to try and decide on the aims of society; its next to define the nature of our existing difficulties, not only material, but intellectual. Only then could it begin to make a rational plan of how to unravel the knots into which we have tied our lives (Waddington, 1948, p. 106).

Such thoughts found their way into 'scientific management', which was believed to avoid the traditional conflicts of worker and boss. A similar process was anticipated in politics. As answers were found to economic and social problems, so political debate would be less divisive. There would be agreement about the need to generate wealth and the means most suitable for achieving this. Political disputes were to be confined to the distribution of that wealth. Science and technology were both the symbol and cause of this new order.

While belief in the virtues of scientific management may have faded, and while the political consensus may have cracked, there has remained an enduring liberal conviction in the neutrality of technology and its general capacity to solve the problems, and meet the ends, of modern society. It can be detected in the expectation that a cure will be found for AIDS, that catalytic converters in cars will end air pollution, and so on. This belief is most apparent in the celebration of contemporary mass culture and the postmodern era it is said to herald (Featherstone, 1988). Technology is merely a benign facilitator of the new order, furnishing us with images and networks out of which we fashion our own vision of reality.

From Marx to the postmodernists, we can detect a consistent faith in technology. It is a corollary of modern society, it is progressive and it holds the solution to many of the problems that afflict our society. The implication is that, rather than curbing the development of technology to protect democracy (as the Greens suggest), we should exploit it fully to preserve democracy. To examine the coherence of this technocentric answer, this chapter focuses on a single case study. We consider the growing support for the idea that the practice of democracy can be enhanced by the application of information technology. In direct contrast to the Greens' concern with the problems that technology causes for democracy, we analyse the case of those who advocate an 'electronic democracy'.

A CASE OF THE TECHNICAL FIX: ELECTRONIC DEMOCRACY

Electronic democracy is becoming an increasingly familiar feature of contemporary political theorising, but not all theorists agree on its desirability. In *Strong Democracy*, Benjamin Barber (1984, p. 274) writes hopefully of 'modern telecommunications technology' being 'developed as an instrument for democratic discourse'. Meanwhile, Michael Walzer (1985, pp. 306–7) describes the spectacle of 'push-button referenda' as 'a false and ultimately degrading way of sharing in the making of decisions'. That the same technology can provoke such different interpretations suggests that the issue deserves closer attention, especially if the stakes are as high as those identified by Barber and Walzer. Barber represents the technical fix approach, and it is around his general arguments that this case study is organised. We begin, though, by looking at the technology itself.

The technology

The technology of electronic democracy is becoming increasingly familiar. It includes computer terminals linked to mainframe databases through the telephone network, cable television with its capacity for 'narrowcasting', and satellite links for multi-channel, international telecommunication. What each of these technologies provides is a cheaper, faster and more sophisticated means of handling information. The key to the development of information technology has been the ability to convert different forms of communication into a single medium – an electronic pulse. Think of how the simple business of writing a letter has been transformed in the last decades. It is no longer necessary to write your thoughts onto paper, address and stamp an envelope, deposit this into a letter box and wait a day or two for it to arrive. Instead, with the aid of electronic mail it is possible for a letter to be written and to be read almost simultaneously. Or alternatively, a letter written conventionally can be FAXed to the recipient within seconds. These applications have been made possible by the ability to use a single medium for the inscription, transmission and reception of communications.

Another possibility realised by the development of information technology, and its conversion of data into a single, electronic medium, is the ability to retrieve, collate, manipulate and correlate vast amounts of data from a variety of sources. When data was stored on paper or cards, such exercises were so complex or time consuming as to be impossible. The application of the new technology not only allows for immense speed in the searching of records, it also eliminates the barriers normally established by space. Records held in different places, in different countries, can be correlated through a central system.

Information technology makes possible the creation of electronic experts. Fallible and forgetful humans can be replaced by interactive data sources. Patients can conduct their own diagnosis by inputting their symptoms in response to a series of prompts. This information can then be set against a complete record of current medical knowledge. In 1991, an advertisement for nurses boasted that they would be aided by an expert system:

> The piece of equipment you see here is just as important as an electrocardiogram or a heart monitor. It's a sophisticated information system specially programmed for nurses' use and it's playing an increasingly vital role in assuring the highest standards for patient care.

This expert system is just one of the developments made possible by information technology. When combined with the other features of information technology – the capacity to communicate and handle information rapidly, and to break the conventions of time and space – it offers the technical basis for electronic democracy.

Electronic democracy

The notion of electronic democracy comes in a variety of guises. The most familiar version (in theory if not in practice) is direct input public opinion polling or voting. This system, which has been used on a large scale in Hawaii and on a smaller scale in Columbus, Ohio, allows individuals to register their views via cable TV networks. Another version of electronic democracy is provided by the expert system or data bank. This enables (or would enable) citizens to have direct access, through a computer terminal, to information about institutions and issues within their society. Citizens are able to interrogate the data bases to establish what is being done by whom.

Electronic democracy is, however, not confined to changing the opportunities available to citizens. It can also be used by those who seek power. The technology may be used by prospective or existing political representatives to discover what package of proposals are most likely to ensure his or her reelection. It can also be used by pressure groups who, by employing direct-mailing lists and cable TV, can target particular constituencies. Mailing lists can help identify a group of individuals with similar interests or values. Such groups can be crossmatched with other lists, to create a constituency which then can be reached and appealed to by the prospective candidate. Indeed, the candidate can devise a set of policies which meet exactly the requirements of each constituency, rather than, as in the past, offering a broad-based platform which was aimed at attracting an undifferentiated electorate. This possibility becomes even more feasible with the proliferation of radio and TV channels. Broadcasters, in building audiences, tend to fragment the viewers/listeners into smaller groups as their interests become more precisely defined. In these circumstances, the citizen is more easily targeted by politicians and interest groups. In short, the technical possibilities introduced by information technology allow for new ways of managing old political tasks – voting, lobbying – as well as establishing new political practices.

The key to these changes is the effect of technology on information. Anthony Downs (1957, Ch. 11) has pointed out that the availability of information can have a decisive impact on the operation of democracy. In an ideal democracy in which everyone has perfect information, the government will accurately reflect people's wishes because it will know exactly what they want, and they in turn will be able to judge the government because they will know everything about its performance. However, in a large complex society, perfect information is not possible. Instead information is costly to acquire and the costs have to be traded-off against the benefits of acquiring it. This creates certain biases within the system and affects the way citizens and politicians in a democracy will behave. Interest groups, with a particular cause to promote and the resources to sustain their participation, will provide free information to political parties and citizens. For the voter, this alleviates the need to engage in expensive information acquisition, particularly where the benefits of voting one way or the other are very small, given the number of other voters. The citizen, therefore, will tend to remain relatively ignorant and the democracy will fail to reflect the people's true interests, only the interests of those who supply useful, cheap information. If voters take only a limited interest in what political parties have to offer, the parties tend to evolve crudely drawn ideologies or images with which to sell their policy package. Images and ideologies convey information in a relatively easy (and free) form. The party's agenda will be set by those who can persuade it that there are interests that ought to be represented. Such interest groups use their resources and the views of their subscribing members to supply information about 'public opinion' to the party leadership.

If Downs is right about the effect of the distribution of, and access to, information in a democracy, then a technology that has a major impact on that distribution (by reducing costs and increasing availability) will itself have a significant effect upon democracy. The question is how this impact should be judged: does it enhance or undermine that democracy?

The case for electronic democracy

Any discussion involving the term 'democracy' inevitably begins with its definition. For our purposes, we need only to acknowledge the distinction drawn between liberal (or elite) and direct (or participatory) democracy. For David Miller, liberal democracy 'is essentially a protective device, a mechanism which obliges rulers to pursue policies in line with the wishes of their subjects'. By contrast, direct democracy 'is essentially a means whereby the people's will is translated into public policy' (Miller, 1983, p. 133). These differences of principle are mirrored in the practices employed. The key mechanism for the operation of liberal democracy is the election; for direct democracy more active participation is envisaged. Advocates of electronic democracy tend to see it as enhancing the latter form of democracy.

One of the most familiar complaints against direct democracy is that it is impractical. Its impracticality is linked to three issues: size, time and knowledge.

Size.   While it may be possible to operate a direct democracy under the conditions that prevailed in ancient Athens with its politically restricted populus, it is impossible to organise mass participation in a large, complex modern state. Such thoughts lie behind Rousseau's (1968) suggestion that the viability of social contractarian democracy depends upon the existence of a small state. They also inform current advocacy of local/communitarian democracy. While there may be no agreement as to what constitutes 'small', it is agreed that once the citizenry reaches a certain size, direct political participation becomes self-defeating and has to be replaced by delegation or representation.

Time. Limits of time impose a similar restriction on participation. Direct involvement in political decision-making is time-consuming if it requires regular attendance at meetings or extensive background reading and research. Most famously, in his *Essay on Government*, James Mill (1937) argued that, in taking up time, political participation prevented individuals from pursuing those activities which benefited both them and society. Lengthy meetings were a waste of time, time which could be better spent on productive labour. Time may restrict participation in another way – where speedy political decisions have to be reached at times of crisis, it is not feasible to consult all concerned.

Knowledge. Given the time constraint, it follows that if citizens are restricted in the time they can allocate to politics, they will be less well-informed, and therefore less able to judge the merits of the policy options available. This was the claim made by Joseph Schumpeter when he argued that lack of direct knowledge of an issue inclined an individual to irrationality. Only when 'national issues' concern citizens 'directly and unmistakenly' do they 'evoke volitions that are genuine and definite enough'. But this is not the typical case: 'Normally, the great political questions take their place in the psychic economy of the typical citizen with those leisure-hour interests that have not attained the rank of hobbies, and with the subjects of irresponsible conversation' (Schumpeter, 1976, pp. 260–1). Even advocates of participation, like J. S. Mill (1972), acknowledge that without direct experience of an issue the capacity of individuals to make coherent judgements is seriously impaired.

Each of these standard objections to participatory democracy is touched on by electronic democracy. As Iain Mclean (1986, p. 141) observes: 'Opponents of direct democracy often dismiss it with a wave of the hand: "It can't be done" .... Now that IT opens possibilities that were unsuspected only a few years ago, this will no longer do.' The size objection, for example, is weakened, according to Barber (1984, pp. 273–4), 'because scale is in part a function of communication' and 'the electronic enhancement of communication offers possible solutions to the dilemmas of scale'. Constraints of time may also be reduced by the ease of access established by information technology – the relevant data can be retrieved from central data banks. Likewise problems concerning knowledge are mitigated because the citizen is able to acquire information through interactive use of data banks and cross-examination of responsible agents and agencies. Insofar as objections to direct democracy are couched in terms of size, time and knowledge, then information technology serves to alleviate each of them. But, of course, there are other reasons for doubting the feasibility of direct democracy.

One of the underlying assumptions of direct democracy is that there is a common interest, and that as a result of popular participation a consensus should emerge about that interest. Direct democracy is not, in this sense, about registering a set of discrete individual or group interests, but rather it is about establishing collective goods and realising them. This feature of direct democracy causes Jane Mansbridge to label it 'unitary democracy'. But, she observes, it is not possible to realise such a democracy in a large society. She explains:

> The participants in a large polity may never meet, and if they do, they will usually know each other only in one role, often one that dramatizes conflicts of interest. Large-scale organisation also requires a hierarchy of some sort, if only for communication. Finally, sheer numbers make impossible a face-to-face meeting of all at once. For these and other reasons, unitary democracy has had no large-scale form (Mansbridge, 1980, pp. 12–13).

But do such reasons still hold under the conditions promised by information technology? We have already discussed how information technology might alleviate the constraints of size. We have also seen how it might be deployed to provide a proxy for more direct contact between citizens, as well as to represent the full range of a citizen's interests. But it can also mitigate the problems of hierarchy. The extensive application of computers is alleged to break down formal structures in the access to information; instead of hierarchies structuring the flow of information, channelling it upwards, making it more exclusive, the dominant form becomes the network. For

Bolter, the computer transforms the relationship between individuals and available knowledge:

> For the technique by which the computer organizes knowledge – nuggets of information linked together into a complex and flexible structure – would make a data base of humanistic literature very different from a library. It would allow the reader to treat written knowledge as a vast workspace in which to build his own interpretive structures (Bolter, 1986, p. 230).

Bolter goes on to suggest that working with computers also demands collaborative, team-based practices. In short, if we accept such claims, with information technology the conditions for unitary direct democracy become possible again. Elitist hierarchies are replaced by democratic networks.

The removal of practical barriers to direct democracy, however, does not exhaust the way in which information technology can be employed to enhance popular rule. Supporters of pluralism and elite forms of democracy can also find applications for it. The re-examination of pluralism by its best advocates, people such as Robert Dahl and Charles Lindblom, has focused on the problems of securing rough equality of access for interests in society. One source of inequality lies in the cost of exercising influence. The new communications technology can reduce the difficulties of coordinating pressure politics. It can cheapen information and therefore put it within the reach of under-resourced groups. At the same time, the opportunities for direct mailing and narrowcasting encourage prospective political representatives to build political programmes which reflect all the interests in their constituency.

Information technology can also address the problems of group formation. For some groups and some issues, as Mancur Olson (1971) has observed, the incentive to mobilise is weak. Where no special benefits are on offer and the costs of action are high, the groups that form will lack cohesion and strength. The explanation for this has less to do with the availability of resources or access, and more to do with the incentives which are needed for group action and the kind of issues with which the political system can deal. Where the goods being sought are public (that is, the benefits are not exclusive to those who campaigned for them or who subscribe to the organisation negotiating for them), there is a disincentive to spend valuable time and money in pursuing the goods. This is rational for all potential participants. As a result, little pressure is exerted and no goods are delivered. The problem is particularly acute for groups with limited resources – the old, the unemployed, the poor. Information technology can, though, aid the coordination of such individuals and reduce the cost of participation. Without altering the overall structure of the individual's perspective, it can reduce the costs of involvement and thereby make political activity more feasible. Such possibilities were important to the European Community discussion of electronic democracy in the early 1980s (Lloyd, 1983).

Information technology also promises to act upon the skewing of interests which occurs in pluralist systems. Downs points out that there is a tendency for producer interests to be over-represented at the expense of consumer interests. Not only are producer groups likely to command greater resources and maintain members' loyalty with the distribution of selective benefits (discounts, advice, social facilities, and so on); they also represent the key site of people's interests and therefore are likely to influence their political behaviour (and the response of their representatives) (Downs, 1957, pp. 253–7). A cause of this bias is the cost and benefits of information about how interests are affected. It is much easier to see how your interests are affected by the closure of the car factory in which you work than it is to discover how the health of your family or the environment generally will be affected by emissions from the cars you build. Any means which can reduce the cost of information will help balance the competing pressures on your actions.

The net effect of information technology on pluralism, therefore, is to erode (further) the party, and to undermine the power of political activists or party paymasters who create political agendas at odds with the popular will. Indeed, the possibilities of information technology have led some political scientists to claim that the technology provides a genuine opportunity to alter the balance of power in democracies. Writing of their experience of 'televoting' in Hawaii, Becker and Slaton claim that

In any polity where governments receive uncertain electoral 'mandates', where public preferences are often unknown or of little interest to those in power, and widespread and growing dissatisfaction prevails with the available opportunities for political expression, opinion polling based on the Hawaii *Televote* approach can play a vital role in restoring genuine public control over their government and their community's political future (Becker and Slaton, 1981, p. 65).

Not all advocates of electronic democracy are quite so messianic in their enthusiasm. McLean (1986 and 1989) sees information technology as offering practical solutions to some of the problems posed by democracy, but he is wary of offering it as a universal panacea. Many theoretical difficulties cannot be subjected to a technical fix. Condorcet's paradox, for instance, which demonstrates the problems of achieving a genuine majority when three options exist, cannot be removed by the application of technology. But even McLean's cautious embrace of information technology is at odds with those who oppose the very idea of electronic democracy.

The case against electronic democracy

The case against electronic democracy addresses both the technology and the underlying account of democracy. The claim is that an exaggerated respect for the technology is used to reinforce a misunderstanding of the nature and practice of democracy. Just because it is possible to organise a push-button referendum, it does not automatically follow that this is what ought to be done.

Roszak complains that advocates of electronic democracy have been seduced by 'the mystique of scientific expertise', and have lost sight of the fact that, 'in a vital democracy, it is not the quantity but the quality of information that matters' (1986, pp. 188–90). A prior question has to be addressed: what model of democracy is appropriate? The technology cannot decide this. It may make direct democracy more feasible; it does not make it politically desirable.

The main arguments about information technology and democracy are, therefore, to be found among those who share a conception of democracy, but disagree about the value of information technology to it. Jean Bethke Elshtain argues that information technology actually undermines the capacity of citizens to perform their democratic function. Far from increasing a citizen's capacity for participation, the electronic plebiscite prevents the 'deliberative process' that is the mark of 'real democracy'. The technology effectively constrains the citizen's capacity for reflecting, creating 'the privatized viewer rather than the public citizen'. Such viewers live in domestic isolation and are unable to interact with other citizens, and therefore, do not develop a sense of 'moral responsibility for one's society and the enhancement of individual possibilities through action in, and for, *res publica*' (Elshtain, 1982, pp. 108–10). Winner too argues that democratic politics requires meeting together in some public exchange, and that this is not the same as 'logging into one's computer, receiving the latest information and sending back an instantaneous digitized response' (Winner, 1986, p. 111). Such a citizen may be more liable to see only a very small world, resulting in 'a public opinion without "public contents"' (Sartori, 1989, p. 53).

Advocates of electronic democracy ally information with judgement. The citizen of the electronic era may be inundated with information, but this does not help with the assessments that are at the core of the democratic process. Winner casts doubt, in contrast to Downs, on the relevance of information to democracy: 'democracy is not founded solely (or even primarily) upon conditions that affect the availability of information. What distinguishes it from other political forms is a recognition that the people as a whole are capable of self-government and that they have a rightful claim to rule' (Winner, 1986, p. 110). Spreading information has no necessary relationship to enhancing democracy.

Private citizens, subjected to vast influxes of information, may in fact have their power reduced. Roszak suggests that, in fact, the 'data glut' provided by information technology actually serves as a form of political control; citizens are overwhelmed by information (Roszak, 1986, p. 188). It is often supposed that crowds, the masses, are the typical object of manipulation. This, at least, is the argument of writers like Schumpeter (1976). A similar thought underpins critiques of various forms of 'mass culture' (Bloom, 1987; Jameson, 1985). But if there is any coherence to the notion of manipulation, then it is plausible to claim that it works most effectively on the isolated individual who is unable to test his/her reaction against that of others, rather than on the masses united in their collective pleasure. Individual citizens who are dependent on information technology for their political participation become potential victims of the abuse of information. It is a process which critics of electronic democracy already see at work.

Politics is becoming a branch of mass marketing; and political participation, rather than being a form of self-development or expression, is becoming an instrument by which tastes are registered and demands made. Nicholas Garnham argues that

> political communication which is forced to channel itself via commercial media . . . becomes the politics of consumerism. Politicians relate to potential voters not as rational beings concerned for the public good, but in the mode of advertising, as creatures of passing and largely irrational appetite, to whose self-interest they must appeal (Garnham, 1986, pp. 47–8).

In summary, critics of electronic democracy resist the technical fix because information technology tends to create conditions which threaten rather than enhance democracy.

The argument about electronic democracy is not, then, simply about points of principle. While democracy is often taken to be about ideas, it is also about the context and conditions that give effect to those ideas. David Lyon draws attention to the fact that, for all the promise and propaganda, experiments in electronic democracy (and the related technology) have met with limited success. These practical inadequacies have been overlaid by the danger that information technology, far from extending the opportunity for participation, may further distort the imbalances within society. 'Political participation', Lyon (1988, p. 93) writes, 'could continue to be the preserve of those with education and means (that is, those who could afford the right equipment and have the expertise to use it)'. These thoughts lead to the view that information technology is more likely to be used to thwart democracy than to enhance it. Citizens may be doubly weakened. The combination of data-banks and electronic surveillance provides the machinery for extending the state's internal power over those citizens. At the same time, the political economy of information technology undermines the sovereignty of states and reduces the significance of state borders, thereby limiting the effectiveness of existing political institutions.

The potential of the technology cannot be separated from the context in which it is located. This is the greatest weakness in the technical fix approach. It treats technology as an instrument, when in fact it involves a complex set of interests. It is significant that in one of the most comprehensive studies of the possibilities and problems of electronic democracy, Arterton's *Teledemocracy* (1987), very little is said about the commercial structure of information technology; that is, the conditions and motives which currently fuel the spread of the technology. As Garnham observes:

> If we see media structures as central to the democratic polity and if the universalism of the one must match that of the latter, clearly the current process by which national media control is being undercut is part of the process by which power is being transferred in the economy to the international level without the parallel development of adequate political or communications structures (Garnham, 1986, p. 52).

For Garnham, democracy in these new conditions depends upon a commitment to 'a public service model of public communication', and it is this principle which must guide the technology, rather than vice versa.

For the critics of the technical fix, the argument for electronic democracy cannot be separated from a discussion of how the technology itself is both organised and regulated. The design of the technology, as well as the rules regulating access to it, become part of the argument as to what democracy involves. Under a system of deregulated satellite TV channels, it is likely that each user/consumer will be encouraged to buy reception dishes. In doing so, they will incur the extra cost of having the dish trained on a particular satellite; different satellites will require new settings; and new channels may expect additional technology (to decode the signal, if it is a pay-as-you-watch service). This particular arrangement is far less conducive to mass participation than would be a system in which there was a single main reception point from which signals could be fed by cable to each home.

The wider context of electronic democracy needs to be broadened in one other direction. The technical fix solution tends to assume that the argument is about the organisation of democracy only. The citizens' preferences are treated as given; the question is merely one of deciding how best to channel and aggregate their choices. It is conceivable, however, that the deployment of electronic democracy will itself alter the character and content of those preferences. While the degree of manipulation is much debated, the fact of the argument means it must be incorporated into any proper assessment of the contribution of information technology to democracy. If the structure of communication changes what is communicated, then theories of democracy cannot afford to overlook this process.

Once again we find ourselves back at the themes which have been the primary concern of this book: the theoretical and practical problems raised by linking democracy and technology. Information technology serves as a further example of, first, how the conditions of political life may be altered by technology, and second, how the form of that technology depends upon political processes, values and choices. No technology, however pervasive or sophisticated, can substitute for politics. The worry lies, however, in the belief that technology can do precisely this. In its most extreme form, this becomes the myth of artificial intelligence and the idea that computers can make qualitatively better decisions than people can. Robert Jastrow, a NASA scientist, once wrote of how the computer will one day supersede humankind. The computer, after all, does not need to suffer the 'wiring defect' which afflicts the human brain (Jastrow, 1978, p. 53). Were this logic to be accepted, democracy would become redundant as a decision method. The next chapter assumes that this logic is flawed, or rather that the decision to secede to technology remains, ultimately, our choice. But before exploring what this means in practice, it is important to return to the idea of the technical fix.

CONCLUSION

The idea that technology holds the answer to social progress is premissed primarily on two claims: (1) that it enables existing tasks to be completed more efficiently; and (2) that it makes possible tasks that otherwise would have been impractical.

Both were clearly involved in our case study. Computers made it possible to organise political participation that would otherwise have not been feasible. Exactly the same reasoning can be detected in Adam Smith's famous account of the benefits of the division of labour. Book One of *The Wealth of Nations* opens with the declaration that: 'the greatest improvement in the productive powers of labour, and the greater part of the skill, dexterity and judgement with which it is anywhere directed, or applied, seem to have been the effects of the division of labour'. 'The division of labour', Smith concludes, 'occasions, in every art, a proportionable increase of the productive powers of labour'. Technology is intimately connected with these benefits: 'the invention of all those machines by which labour is so much facilitated and abridged seems to have been originally owing to the division of labour' (Smith, 1986, pp. 109–14). It is important to note that the value of technology lies in its application to pre-existing practices or imagined ends. In other words, it is an *instrument*; it does not add anything except speed, convenience and economy to those practices or goals which already feature in society. Technology may help to solve problems, but those problems are posed by, or spring from, prior political decisions and choices.

The failure to understand the instrumental role played by technology can open the door to political oppression. If technology is presented as an author rather than as a instrument, it can be used to eliminate political argument. Adorno and Horkheimer (1979, p. 121), for instance, claim that 'a technological rationale is the rationale of domination itself. It is the coercive nature of society alienated from itself. Automobiles, bombs, and movies keep the whole thing together until their levelling element shows its strength in the very wrong which it furthered'. Whether or not we accept the particular purpose attributed to technology here, the central claim is that the use of technology is not neutral; that it actually shapes the ends it is alleged to be serving.

Critics of the technical fix question the idea that efficiency is morally neutral, that it represents an objective criterion of assessment and that this justifies the spread of technology. For the critic, terms such as 'efficiency' rest upon a loaded understanding of how social relations *ought* to be ordered. Alasdair Macintyre links efficiency to effectiveness, and argues that 'the whole concept of effectiveness is inseparable from a mode of human existence in which the contrivance of means is in central part the manipulation of human beings into compliant patterns of behaviour' (Macintyre, 1981, p. 71). Whenever words like efficiency or effective are used, they beg questions as to what end is being sought and how we are to assess the relative efficiency/effectiveness of any one method over another. Any use of the word efficiency or superiority in respect of technology must be followed by questioning why such reasoning was compelling – what counts as 'superior' (MacKenzie, 1990, p. 11). It may even be that efficiency is not the unalloyed good that the technical fix supposes. *New Republic* (18 December 1990) once argued that the way in which

> information technologies, including television, have damaged American democracy is by making it more democratic. The people's will is being translated into electoral discourse and public policy with record efficiency – more efficiency than the Founding Fathers intended, and more efficiency than is good for us.

The same thought occurs when we read that the electronics firm Pioneer are offering, for a mere £1250, a device which uses computers, satellites, liquid crystal displays and much else besides, to enable (very rich) car drivers to find out where they are. What is wrong, you might ask, with a map?

In the context of the application of technology to democracy, these arguments highlight two objections to 'electronic democracy'. The first focuses upon the model of democracy which is being made more efficient. The second exposes the methods being used to enhance this 'efficiency'.

The technology's ostensible use as an instrument disguises its role in reinforcing a set of views or values which may not have been freely chosen, but were in fact imposed. Or more subtly the technology, instead of being a neutral instrument, may actually induce a particular set of preferences in those who use it. Rather than using the technology, people are used by it.

To see technology as operating like this automatically calls into question the idea of 'the technical fix'. Technology cannot be seen simply as a technique, an instrument of some external purpose. Instead, it has to be viewed as something which introduces a change in the relationships which existed previously. If such influence is seen as inherently malign, then we find ourselves once again contemplating the Green solution. But as we suggested in Chapter 7, this may not be a viable route to follow. The indication from these last two chapters is, rather, that we have to find a way of analysing and utilising technology which neither treats it as holding the key to all our problems, nor on the other, requires the renunciation of much current technology in the name of democratic control. Instead, we need to develop a way of integrating technology with political analysis.

9 Democracy and Technology

INTRODUCTION

In March 1991, Britain was swept by a wave of public reactions, ranging from enthusiasm to condemnation, to a news story about 'virgin birth'. Public attention was drawn to the fact that women could conceive children without sexual intercourse. There was nothing new about this possibility; indeed the technology had been available and applied for some time. But this was the first time 'public opinion' had found a voice. Captured in the rhetoric that surrounded the artificial insemination of a single woman who had never had sexual intercourse were a great number of very deep-seated beliefs about sexuality and sex, marriage and families. This was not just an argument about a technology, it was about competing ways of life and opposing value systems. The capacity of technology to provoke this kind of reaction is what lies behind this chapter.

The last two chapters have examined two very different attempts to reconcile democracy and technology. The Greens' solution to the problem is to subject the technology to their political principles. The technical fix answer is to allow the technology to lead the political order. Although both approaches provide important insights into the relationship between politics and technology, neither response is entirely satisfactory. In this chapter, therefore, we explore a third means of combining technology and democracy.

The strength of the Green position lies in the political emphasis it puts upon the impact and role of technology. In particular, the Greens highlight the need to connect political principles to technological practices, to see how technology can reflect certain political values while appearing to exclude others. Thus it is that technology can be labelled as appropriate or inappropriate.

The key weaknesses in the Greens' position derives, first, from the political principles to which they lay claim. These are derived from an idea of 'nature'. The problem is that their 'nature' owes more to particular preconceptions and cultural norms than to any firm features of the natural world. There is no persuasive reason for accepting the Greens' account of our duties to nature. Equally suspect are the principles which organise the Greens' account of democracy. They seek to marry the local and the global, but it is not clear that local actions will generate global solutions. This problem is compounded by the Greens' emphasis on cultural change. It is not clear how altering attitudes and expectations could influence a shift in the location and exercise of power. In their emphasis on their political principles, the Greens overlook the structures which tie technology to dominant interests and the existing order. The technical fix argument recognises and embraces the inextricable bond with technology. It acknowledges the impossibility of making technology subservient to political principle. But in doing this, it attributes to the technology a purpose and rationale of its own, ignoring the impact of political interests and values upon the form and effect of that technology. There is a refusal to see the contextual dimensions to technology.

Although they differ in their weaknesses and their strengths, the two arguments share a common flaw. Both create a false dichotomy between politics and technology. The Greens want to recreate a pure politics in which 'human' values can survive and flourish. For this to occur technology must first be reduced to a subservient position. The technical fix school, on the other hand, want to eliminate the 'irrationalities' of politics through the deployment of technological progress. While the two groups have drawn opposite conclusions, their arguments rest upon a common assumption: that politics and technology can be separated. But such a separation is impossible to make.

Technology cannot be parted from political processes, and politics cannot be analysed independently of technology. Let us consider the first part of this claim. In a brilliant study of the development of missile guidance systems, Donald MacKenzie shows how the technology was shaped by social forces. The 'facts' of missile guidance, in particular the notion of 'accuracy', were not just there to be observed; they had to be created. Ideas of accuracy were not simply technical descriptions of the elimination of error. They were actually 'the product of a complex process of conflict and collaboration between a range of social actors including ambitious, energetic technologists, laboratories and corporations, and political and military leaders and the organisations they head' (MacKenzie, 1990, p. 3). These conflicts arose over the definition and measurement of accuracy, a term for which there could be no absolute meaning. This uncertainty created the forum for the political processes: 'The same premise, uncertainty of accuracy, was drawn on to argue for manned bombers, for the cancellation of MX and a more "dovish" defense policy, for radio guidance and for larger missile warheads!' (ibid., p. 363). Arguments over accuracy were not confined to competing interpretations; the 'facts' themselves were socially constructed. The calculation of accuracy, though coded in detailed mathematical models, was rich with assumptions; 'nowhere in this complex process of modelling and testing did unchallengeable, elementary "atomic" facts exist' (ibid., p. 381).

Such insights into the integration of politics and technology are not exclusive to the contemporary world. They apply equally to the emergence of the bicycle (Pinch and Bijker, 1987). The technology, and the 'facts' upon which it rests, result from a process of social construction. Analyses of this kind endorse the idea that technology cannot be separated from politics. But there is a second dimension to the claim being advanced here. It is not just that politics shapes technology, but that technology shapes politics. MacKenzie concludes his study of nuclear weapons: 'technology in the nuclear world is not above politics as an autonomous determining factor, nor beneath it as a dependent effect, but part of it' (ibid., p. 412). What is true for the nuclear world is true for other technologies.

It is impossible, I want to suggest, to think of politics without reference to technology. What we conceive of as 'human' or 'natural' is more often than not the creation of our technology. The way we see and experience the world, the way we come to define our place within it, all of these are achieved by and through forms of technology. At the same time, the technology we employ and the ends it is intended to serve are themselves shaped by political processes. This point is powerfully made by Judy Wajcman in her feminist analysis of technology. She argues that technology is a form of culture through which male gender identity is constituted (Wajcman, 1991, p. 158).

For Jurgen Habermas, the link between technology and politics is founded in human nature itself:

At first the functions of the motor apparatus (hands and legs) were augmented and replaced, followed by energy production (of the human body), the functions of the sensory apparatus (ears, eyes, and skin), and finally by the functions of the governing centre (the brain). Technological development thus follows a logic that corresponds to the structure of purposive–rational action regulated by its own results, which is in fact the structure of *work*. Realizing this, it is impossible to envisage how, as long as the organization of human nature does not change and as long therefore as we have to achieve self-preservation through social labour and with the aid of means that substitute for work, we could renounce technology, more particularly *our* technology, in favour of a qualitatively different one (Habermas, 1971, p. 86; his emphasis).

Habermas is, therefore, deeply sceptical of attempts to do away with technology in general; it is too closely allied to how we think and act. At the same time, he is not committed to seeing all forms of technology as the logical result of human needs. Technology may develop inappropriate forms and thwart existing social practice; in particular it may come to substitute for the rational–purposive action which drives his conception of human nature. Rational–purposive action is what establishes the goals and principles which order society and which technology should serve. The problem is establishing when this relationship breaks down, and when technology comes to dominate. The answer cannot derive from some simple assertion of what is 'human', since what this means is inextricably linked to the technology. Establishing what political principles or values are in operation, or whose interests a technology serves, must depend upon a close examination of the technology itself, and the relations which form it and are formed by it.

THE CULTURE OF TECHNOLOGY

This alternative perspective portrays a much less certain world than we find in either the ecocentric or the technocentric perspectives. It is borrowed from the arguments that emerged in Chapter 2 about the contingent nature of technical change and its relationship with politics, but it also draws on much else in this book. It builds upon the earlier discussion of the political generation of technology (Chapters 3 and 4), of the political effects of technology (Chapter 5) and of the problems encountered in choosing between technologies (Chapter 6).

The alternative perspective, which I label the 'cultural approach' to technology, seeks to integrate political argument and technological assessment. In doing so, it follows the critiques of science and technology to be found in, among other places, feminist writing. The study of technology cannot be separated from the surrounding political structure. So in the case of feminism, the argument takes this form: 'If women are to transform or "re-valence" technology, we must develop ways to assess the equity implications of technological development and develop strategies for changing social relationships as well as mechanical techniques' (Bush, 1983, p. 163). The implications of this are that technology is a 'human cultural activity', and therefore, 'the challenge to feminists is to transform society in order to make technology equitable and to transform technology in order to make society equitable' (Bush, 1983, pp. 164–8).

A similar approach is advocated by Schwarz and Thompson who also refuse the technology–politics dichotomy, and who choose to see technology as 'a social process'. The development and assessment of technology, from this perspective, contains an important cultural element, by which contending understandings and value systems struggle to impose their meanings and interpretations upon a technology. Technological development is 'the product of an untidy process of *bricolage*, never of an immaculate, surprise-free, and fully specified scheme' (Schwarz and Thompson, 1990, p. 144). The assessment of technology requires taking part in an argument about competing cultural and political values. It is not an argument that can only be about the advertisers' and manufacturers' blurb.

This general approach also underpins Winner's use of Wittgenstein's 'forms of life'. Winner wants to capture the way in which technology inhabits 'the texture of everyday existence': 'the devices, techniques and systems we adopt shed their tool-like qualities to become part of our very humanity' (Winner, 1986, p. 12). For Winner, the implication of this is that technical change has to be accompanied by a series of questions about more than 'the physical instruments and processes . . . but also the production of psychological, social, and political conditions' (ibid., p. 17). According to this perspective, we are neither the creatures nor the controllers of our technology; we are caught up in it, simultaneously directing it and coping with it.

While such approaches constitute important rivals to the stark dichotomies of both the technical fix and the Green visions, they still leave some difficult questions to answer. While it is helpful to see the cultural and other dimensions to technology, we need to go beyond these general statements of intent, to see how different technological systems work in shaping or expressing political values. Technology cannot easily be labelled 'good' or 'bad', not just because much depends on how it is used, regulated or designed, but because the sort of position that such judgements imply is not available to us. Political principles emerge in conjunction with the technology; they can be neither subservient to it nor completely dominant over it. And the people who reflect upon their relationship with technology are themselves shaped by it. It is not merely that technology enables us to do more things faster and cheaper than before. It is also that in the process who we are and what we want is also changed. We are different people *because* of our technology.

In integrating politics and technology in this way, there is a danger that the wider context gets lost to view. It is not the case that technology is just what we make it, or that we are simply made by our technology. It is that who 'we' are needs careful scrutiny; 'we' do not exist separately from the technology and the politics. As MacKenzie observes, 'changes in technology go hand-in-hand with changes, small and large, in the preconditions of their use, in the way they are used, in who uses them, and in the reasons for their use' (MacKenzie, 1990, p. 9). Technology is frequently employed to express and reinforce existing dominant interests. In adopting a cultural approach to technology, these material factors do not disappear. Culture is no more free of conflict of interests than is any other activity. And the cultural approach to technology does not seek to lift technology away from its material influences, or to deny that it serves and thwarts certain interests, but rather to focus upon *how* this political role is contained within the technology. The cultural approach is an attempt to get away from the view that technology is either an 'instrument' or a 'barrier', something to be used or confronted. Instead technology is a crucial aspect of the way political aspirations and actions are organised. This does not mean technology cannot be judged, but it does mean that analysis of both the technology and the politics cannot proceed along conventional, separatist lines.

Political principles cannot be derived simply from introspection or profound reflection. They must derive in part from our relationship with our technology. As a result we are forced to ask not what we mean by 'democracy', but what does it mean to be a democrat, given our technology? This is not to suggest that we have to accept existing technology as a 'natural fact' about the world we inhabit. It is equally important to ask how that technology might be changed to accommodate our principles, but this questioning begins with the technology as we find it, rather than as we would like it to be.

CASE STUDY: FREEDOM OF SPEECH AND MASS COMMUNICATIONS

There is, of course, no simple – or single – answer to the above questions. Much depends upon a detailed appreciation of the conditions which both created and are created by technology. To keep the task a manageable one, I concentrate only on mass communications technology. The reasons for doing so are perhaps obvious. Not only has mass communications technology changed dramatically and rapidly in recent years, but its impact has been widely felt. As Hans Magnus Enzensberger observed nearly twenty years ago:

> With the development of the electronic media, the industry that shapes consciousness has become the pacemaker for the social and economic development of societies in the late industrial age. It infiltrates into all other sectors of production, takes over more and more directional and control functions, and determines the standard of the prevailing technology (Enzensberger, 1976, p. 20).

There is no doubting the prescience of Enzensberger's claim. The development of satellite and cable networks has further expanded the capacity for transmitting messages; they have also vastly increased the range of signals that can be received. The mobility and flexibility of recording technology has enabled the development of electronic news gathering (ENG), which in turn has affected the speed and reach of TV news coverage. Changes in print technology have drastically reduced the costs of newspaper production, just as desk-top publishing (DTP) technology has expanded the possibilities for individual initiatives. FAX machines and electronic mail have established new channels and forms of communication. Together all these have changed the way information is handled, distributed and used; and in doing this, the way we live and think have been changed. It is with this world that our political principles has to deal; it is within this world that we have to establish the means, ends and conditions of democracy. How then do we link principles and practices? My answer is organised around a number of key ideas within the concept of democracy, using these as starting points from which to discuss the impact of, and possible reforms in, communications technology. First, though, we need to sketch out a definition of democracy to serve as a backdrop to the more detailed focus on certain ideas within it.

Democracy

Any account of democracy works with a number of central ideas in its attempt to give shape to 'rule by the people'. It seeks to establish means by which 'the people' can articulate their values, wishes or desires. These have then to be registered within the political system and converted into a set of policies or procedures, which in turn have to be implemented and assessed. For such a process to work, certain conditions and mechanisms have to exist. The ability to articulate views, for example, depends upon some version of freedom of speech; the expression of these views also requires the presence of political organisations (trade unions, pressure groups, movements, political parties) which can mediate between popular opinion and the political system. In turn, the political system has to be capable of representing or responding to the demands put upon it. There has to be some means by which its authority can be legitimated. All these activities need to be contained within a political culture which nurtures and maintains the values of democratic practice. And finally, there must be some means by which political authority can be checked whilst remaining sufficiently resilient to resist rival illegitimate claims to its power. There is nothing very controversial about this simple model of democracy (although there are many variants upon it), but in all its aspects it is profoundly affected by communications technology; this is especially true of freedom of speech.

Freedom of speech

There is no doubting the importance of freedom of speech to theories of liberal democracy. John Stuart Mill writes in *On Liberty*, the principles of which underpin his account of representative government, that 'if all mankind minus one were of one opinion, and only one person were of the contrary opinion, mankind would be no more justified in silencing that one person, than he, if he had the power, would be justified in silencing mankind' (Mill, 1972, p. 79). Joseph Schumpeter (1976, p. 295), despite his desire to revise 'classical' accounts of democracy, also insists that for his new democracy to work there must be 'a large measure of tolerance for difference of opinion'. Today, such principles are enshrined in documents such as the Universal Declaration of Human Rights: 'Everyone has the right to freedom of opinion and expression; this right includes freedom to hold opinions without interference and to seek, receive and impart information and ideas through any media and regardless of frontiers' (Article 19). The general sentiments of such claims are widely shared, and are an organising principle of groups like Amnesty International and Charter 88.

What is most frustrating about such pronouncements is that virtually nothing is said about the means by which the various goals are to be achieved. Writing about the mass media and democracy, John Keane comments that 'almost nobody asks basic questions about the relationship between democratic ideals and institutions and the contemporary media' (Keane, 1991, p. x). What exactly is meant by 'freedom of speech' in a world in which the means of communication are owned by media conglomerates and in which access to these means is very restricted? 'Whereas the media have changed dramatically over the last two centuries', writes Simon Lee (1990, p. 21), 'the same old arguments over free speech seem to carry on regardless'. Part of the explanation may lie in liberalism's predominant concern with restraining the powers of government. As John Thompson argues: 'By placing so much emphasis on the dangers of state power, the early liberal theorists did not take sufficient account of a threat stemming from a different source: from the unhindered growth of media industries *qua* commercial concerns' (Thompson, 1990, p. 18). This is a serious omission because, as Lee observes (1990, p. 24), 'Those who can buy a newspaper or a television station have far greater access to the so-called free marketplace of ideas than have ordinary citizens'. In short there are two gaps in the link between democracy and the technology of mass communication. The first is the *meaning* of 'freedom of speech' in the context of mass communications; the second is its *operation*.

There is a history to this predicament. As Ithiel de Sola Pool (1983) argued, despite the US Constitution's commitment to protect freedom of speech through the First Amendment, it has found no direct expression in the actual organisation of broadcasting. Instead, the allocation of radio frequencies has been determined by the demands made by radio stations and by the relative weakness of the Federal Communications Commission, the regulatory body. A weakly regulated market cannot guarantee freedom of speech.

In Britain, by contrast, the failure to integrate the principle into practice can also be traced to the origins of broadcasting. Unlike in the US, in Britain the allocation of radio frequencies was carried out very cautiously and protectively. But this caution did not arise from a concern for the principle of free speech; it derived from the military security considerations which accompanied the development of radio technology in Britain (see Chapter 3). Broadcasting was (and still continues to be) treated as an aspect of the state's interests, rather than the people's.

But to observe the origins of this divide between political principles and technical systems is not to reunite them. For that, there needs to be further analysis of the features and functions of the technology. Few attempts have been made. Pool (1983) and Enzensberger (1976) are notable exceptions.

In his *Technologies of Freedom*, Pool sees communications technology as bringing about the conditions of freedom. 'Electronic media . . . allow for more knowledge, easier access, and freer speech than were ever enjoyed before' (1983, p. 251). The only threats to this process are interfering politicians and their bureaucratic routines. The answer lies with the free rein of market forces, albeit framed by the commitment to freedom of speech enshrined in the First Amendment. By contrast, Enzensberger sees electronic media as contributing to freedom only when they are organised in a more deliberative way. The possibility of an 'emancipatory use of media' depends on a number of conditions being fulfilled. These are illustrated by the comparison he draws with the current, 'repressive' mode:

Repressive	*Emancipatory*
Centrally controlled	Decentralized control
One transmitter, many receivers	Every receiver a potential transmitter
Immobilisation of isolated individuals	Mobilization of the masses
Passive consumer behaviour	Interaction of those involved, feedback
Depoliticization	A political learning process
Production by specialists	Collective production
Control by property owners	Social control by self-organization

Source: Enzensberger, 1976, p. 38.

Enzensberger offered these alternatives in the 1960s, but the underlying theme still echoes. Keane, for instance, presents the 'public service media' as an important challenge to the rival claims of 'technocratic solutions': 'Democratic public service media are reflexive means of controlling the exercise of power. They are unsurpassed methods of checking the unending arrogance and foolishness of those who wield it' (Keane, 1991, pp. 179–81). At the same time, he advocates the use of new technology – 'interactive television, digital copiers, camcorders and music synthesizers' – to facilitate communication among citizens (ibid., p. 159).

My immediate concern is not with the particular solutions offered by Pool, Enzensberger and Keane, but with the general categories contained within them. First, there is the *design* of the technology and its potential applications – as receiver and transmitter; secondly, there is the *organisation* of the use of the technology – who has access and control; and finally, there is the *content* which it carries – political education. The general implication of the argument is that the three elements must be included in any account of the politics involved. What follows is an attempt to see what might fall under these headings if democracy-as-freedom-of-speech and mass communications are to be combined.

Organisation

If the principle of freedom of speech is to be applied to mass communications technology, what might this mean? Tannsjo (1985) provides one answer by asking whether mass communications can serve democracy where the right to 'freedom of expression' operates. 'Freedom of expression' is realised in the private right to own the means of mass communication. Tannsjo asks whether this right is compatible with what he describes as 'sound' mass communications. By this he means a situation in which mass communications:

1. 'contribute effectively to the *growth of knowledge* in society';
2. 'allow for a *pluralism of ideas*';
3. provide '*equal opportunity* for various social and cultural groups and strata in society to acquire access to media' (Tannsjo, 1985, pp. 553–4, his emphasis).

Tannsjo argues that none of these conditions can apply under a system of mass communications in which the 'freedom of expression' is also recognised. He argues that the market in mass communications has led to 'a concentration of capital and centralization of ownership'. Although concentration of ownership in a few hands is the key defining feature of the political economy of mass communications, it is also marked by both its transnational reach and the integration of media forms. The major media interests are not concerned with only one form of communication. The days of the press baron are over. Time-Warner, for instance, have interests in film, television, music and magazines. Rupert Murdoch has substantial interests in film, TV (cable and network), books and newspapers and magazines. This does not just represent a broad portfolio but an enclosed network of interests. Time-Warner sell films through the soundtracks they market, and vice versa. Murdoch's newspapers are used to promote the demand for satellite channels, and vice versa.

The new communications order represents an immense concentration of power. The effect of this can be to restrict access to the means of communication *within* conglomerates or to prevent competition *between* them. In such circumstances, Tannsjo concludes (pp. 557–9), 'sound' mass communications can only exist where freedom of expression is *curbed*. Individuals must be prevented from purchasing their own television stations.

What is to replace the old, commercial order? Tannsjo argues, rather weakly, for 'democratic political control', by which he does not mean state censorship but does mean 'strong political interference with the flow of information'. The feebleness of this conclusion is a result of the lack of any real attempt to address the other side of the argument: the content of 'sound' mass communications, the second of Enzensberger's general categories. 'Democratic political control' means nothing unless we also know what must be controlled and what provides control.

Content

In the argument about how democracy can be linked to mass communication, a key assumption is that the mass media – newspapers, radio, TV – are responsible for furnishing us with the political information that shapes our view of the world and of ourselves. As Angus and Jhally (1989, p. 2) write: 'In contemporary culture the media have become central to the constitution of social identity'. Today virtually no one, if the surveys are to be believed, depends upon conversation with their neighbours for political information (Negrine, 1989). It is true that there are variations in the sources upon which people draw. For example, readers of the tabloid press regard television as the most reliable source of political information; whereas readers of the broadsheet press regard their papers as the best source of information. Whatever the variations in people's perception, the standard claim is that mass media furnish us with information about the world.

A second assumption is that this information affects the way we behave, that citizens' political behaviour is powerfully shaped by the information they receive. The collapse of communism in Eastern Europe in 1989–90 was attributed to a wide range of factors, but one in particular stood out. It was the idea that the popular uprising against state socialism had been initiated by Western media coverage. The Deputy Editor of the British *Channel 4 News*, Garron Baines, wrote:

It was a fitting climax to revolution in Eastern Europe that it was played out live to a worldwide television audience from the Romanian TV station that was itself a battleground. . . . Western television has become a catalyst of revolution. Demonstrations and protests, subsequently broadcast by national television inside those countries undergoing change, has accelerated events at a rate only achieved by unfettered and direct communications.

Sir Geoffrey Cox, an ex-editor of the British Independent Television News, was equally convinced of television's revolutionary powers: 'The revolution sweeping Europe is a television revolution. The television cameras have played the part of the trumpets of a modern Jericho'. These extravagant claims for the power of television have to be treated with considerable caution. Television rarely, if ever, has such a dramatic effect, but this should not cause us to discount its influence completely. The question is whether the impact is determined by the content of the broadcast or the context in which it is received, or possibly in the way the original activities are designed to attract coverage?

Television has certainly made available new kinds of political actions. As Thompson observes:

The very existence of the medium of television gives rise to a category or categories of action which is carried out with the aim of being televisable, that is, capable of being regarded as worthy of transmission via television to a spatially distant and potentially vast audience. Today part of the purpose of actions such as mass demonstrations and hijackings, summit meetings and state visits, is to generate televisable events which will enable individuals or groups to communicate with remote and extended audiences (Thompson, 1990, p. 231).

Television, in this sense, has become part of the practice of politics, and therefore has to be incorporated into the principles attached to politics. What access is granted, what is filmed and how, must be incorporated into an analysis of the media's role in a democracy and its impact on political behaviour.

Television does not just provoke actions, it also expresses political ideas. The issue is what such expression involves. The 'speech' within 'freedom of speech' cannot refer simply to the words that are spoken; it must also include images and sounds. Furthermore, it cannot be confined to those areas of the mass media which are formally designated as 'political' – the news, current affairs, and so on. This broad definition of 'speech' is implicitly recognised in current practice. One of the strangest features of the early days of the Gulf war was the way programme controllers in Britain became overwhelmed by a desire to 'pull' shows that were deemed unsuitable for 'a nation at war'. Out went 'Allo 'Allo', a situation comedy set in wartime France, and 'The Clothes Show' dropped its item on current army fashions. Reports appeared in the papers that BBC Radio 1, the pop station, had issued a list of records that were deemed unsuitable. Out was to go Phil Collins' 'In the Air Tonight', Elton John's 'Saturday Night's All Right for Fighting' and Lulu's 'Boom Bang-a-Bang'. Sadly, these reports were denied by the BBC. But whether or not they were true, they were entirely plausible. There is never any shortage of people who will claim that pop music is the source of moral degradation or of political emancipation (Street, 1986). They are rarely right, but in their excess of enthusiasm for the power of popular culture, they have latched onto an important insight. What the programme controllers implicitly recognised in their agonising over suitable programmes is that 'information' is only important when it is used, when it becomes 'knowledge', when it is given a form and purpose which produces some action or reaction.

This is important to our attempt to link democracy and technology. The technical means employed in imparting knowledge actually changes the character of that knowledge. The form of the technology can alter the *kind* of news that is reported. This is not simply measured by the speed at which a report reaches its destination or by the quality of the broadcast sound and pictures. The introduction, for instance, of electronic newsgathering techniques means that it is possible to broadcast 'as things happen'. That, at least, is the rhetoric. The potential of the technology – to be carried easily, to be edited quickly and so on – becomes part of the business of covering the news. The technology is held to present a 'truer' account of events. In fact, the technology introduces a new version of 'events', which may be no more or no less true. Wallis and Baran describe, for example, how the combination of satellite and ENG technology has changed the definition of news:

> The engineer with the ground station and the task of establishing a link becomes the first person at an event, not the correspondent with the task of asking questions about what's happened. With constant feeds from news processing centres to news purchasers . . . *speed* assumes top priority, sometimes even over *investigation* (Wallis and Baran, 1990, pp. 19–20, their emphasis).

Or as they remark elsewhere, 'accuracy enters a risky duel with urgency' (ibid., p. 220). Meanwhile in the studio, the use of computer graphics technology provides new ways of packaging and representing news events.

In reducing the costs and increasing the range of television transmission, direct broadcast satellites have also led to a change in the way the world is perceived. The ease of access to global TV actually allows for the dominance of US culture. And with the constant supply of news provided by channels like CNN, there is less likelihood of resources being committed to local news coverage (ibid., p. 145). Together, these trends create a world with a limited and narrow horizon.

If the technology does change the form of the news and the knowledge it represents, then it is once more implicated in any attempt to analyse the democratic form of mass communications. It becomes necessary to ask how and in what way the technology changes the content of the knowledge being conveyed.

Design

Implicit in the answer to such questions is the role of design in democratic accounting. Technologies can be more or less capable of control by users, more or less drastic in the effects they create. Although design may not be the final determinant of such things, it plays a crucial part. Adorno and Horkheimer implied as much in the language they use to describe technological change:

> The step from the telephone to the radio has clearly distinguished the roles. The former still allowed the subscriber to play the role of subject, and was liberal. The latter is democratic: it turns all participants into listeners and authoritatively subjects them to broadcast programs which are all exactly the same (Adorno and Horkheimer, 1979, pp. 121–2).

In a similar spirit, Rudolph Bahro specifies particular conditions which should be incorporated into a system of mass communications in order for it to meet the conditions of democracy. Included among these is the duty 'to prevent the broad-band cable network and computerized data gathering as instruments of increasing isolation, control and intimidation of the citizen' (Bahro, 1986, p. 41). Ivan Illich's (1975) notion of 'convivial technology' – technology that is user friendly – conveys a parallel idea. In short, the design of the technology, the systems determining access to it, and the use made of it, have to be subjected to political analysis to establish the principles and values incorporated in it.

This case study of mass communications and democracy has examined three aspects of the technology: the way it is organised, how it shapes content, and what is incorporated in its design. Each affects the way it works and what, therefore, must be considered in introducing any form of democratic control. There are, of course, other ways in which the same technology could have been analysed, but the point has been to show what sort of technical questions have to be incorporated into an attempt to link politics and technology. The aim of this case study has been to provide an *approach*, rather than a set of answers. It suggests what ought to be considered in establishing freedom of speech within a system of mass communications; it has said little about the conclusions of any such examination. In the final pages of this book, I want to draw out the implications of this approach for the study and practice of politics.

POLITICAL IMPLICATIONS

There is a temptation to take the cultural perspective too far, to see it as telling the whole story of technical development. This would be wrong, at least if it led to the conclusion either that the technology was not part of wider social conflicts or that the technology itself did not impose limits on the practices and principles to which it gives shape. It would be equally wrong if it denied the relevance of specialised knowledge or of the institutions which assess or regulate the development and application of technology. What the cultural perspective provides is a way of thinking about expertise, regulation and assessment, and a way of producing alternative means of organising such things.

The implications for risk assessment, and decision rules generally, are that moral and cultural precepts need to be incorporated in the judgement about levels of safety and so forth. At the same time, it is not to be supposed that only moral values need be considered, nor that only 'ordinary' citizens should be involved. Where public participation is deemed appropriate, it has to be informed by a broadly-based understanding of the issues and techniques involved. For this to occur, both education and public media have to help provide the capacity for responsible decisions. O'Riordan proposes the development of a 'vernacular science', 'an interactive science of modelling and intuition, of analysis and ethos, of instruction and participation' (O'Riordan, 1991, p. 162).

It is sometimes suggested by those who stress the need for public education in science and technology, and for better media coverage of them, that the end result would be a citizenry capable of rationally assessing their technological options. Nelkin (1987, p. ix), for instance, writes of her 'conviction that fair, critical, and comprehensive reporting about science and technology is extremely important in a society increasingly dependent on technological expertise'. But the implication of the cultural approach is that such a goal is neither wholly feasible nor desirable. Embedded in our response to and relationship with technology are a set of ideas and emotions which cannot be subjected to a straightforward system of technology assessment. We are unable to put a clear distance between ourselves and what we are assessing. The technology of sound recording and music making illustrate this.

Experience of performing and hearing music is invested with meanings which the technology can simultaneously encourage and deny. Goodwin notes how it is thought that analogue recordings are 'warmer and more natural than digital', and that this is reproduced in their respective 'visual signification . . . – waves as opposed to numbers' (Goodwin, 1990, p. 265). It is not clear whether these judgements are either right or wrong, or whether they can be subjected to rational assessment. They are, nonetheless, important components of decisions about technology. And they are not peculiar to the technology of music – no technology is assessed purely in terms of its function.

In drawing out the implications of a number of planning disasters, Hall (1981, pp. 267–72) argues for the need to devote more attention to forecasting. At one level, this simply means greater time and effort being spent on the technical problems of extrapolation and risk accounting. It is, though, not simply a matter of technique; it also requires the recognition that forecasting has to be conducted in conjunction with evaluation, which includes the values of those affected by the technology as well as of those implementing it. A similar idea is captured in Pacey's notion of 'innovative dialogue', by which all parties to a technological decision are in some way involved. Such dialogue, says Pacey (1983, p. 159), provides 'a dialectic between conflicting values' and 'is important as a means of balancing narrow specialist views against broader insights'. Dialogue too plays a vital part in Goldhaber's (1986, pp. 230–1) reforms of technology policy-making.

Such general sentiments, which in one way or another draw upon the cultural account of technology, are of limited significance unless located in a specific set of institutional arrangements. One important implication of the cultural approach to technology is the recognition that there is no one source of knowledge about technology (in contrast to both the technical fix and the Green vision, where knowledge is entrusted to a single key source – to experts and to the community, respectively). It follows, therefore, that centralised decision-making is inappropriate. As Burnheim (1985, p. 3) argues, the complexity and variegated character of technology decisions makes comprehensive understanding impossible, and in such circumstances, decentralisation of decision-making becomes the best means of registering and acknowledging the diversity of insights and perspectives.

The institutions involved, though, cannot be confined to the typical agencies of the state. There is, as we have seen, no one model by which technology emerges, nor one role for the state to play. Drawing conclusions from his study of two major technological 'errors', Henderson (1977, pp. 192–4) argues for pluralism which is to be represented in the range of both information and institutions. Mistakes result, he argues, from reliance upon a single source of information and judgement about a technology. Only when knowledge and responsibility are widely distributed can policy be adequately scrutinised.

These things are, of course, easier to write than to achieve. Ensuring that a technology represents the interests of all those connected with it depends on the way power is distributed around it. Paul Theberge (1990) offers a cautionary tale to counter any naive expectations. His detailed study of the emergence of Musical Instrument Digital Interface (MIDI), a device for allowing keyboards, drum machines and so forth to communicate with each other, shows how user groups tried to establish principles which would make the technology 'democratic'. But in their battle with the manufacturers, these amateur groups were faced with, on the one hand, the rival claims of professional users and, on the other, MIDI makers who wanted to incorporate the groups for marketing purposes.

In the struggle for control of technology, given the importance of design to the political effects of technology, action cannot be directed only at its physical form. The ideas and practices behind that form also require examination. In particular, science must come under scrutiny. One of the lessons of Chapter 4 was that the politics of science are not confined to its application, but are present in both its organisation and disciplines. Lynda Birke (1986), for example, reflects upon the reforms necessary to achieve gender equality in science. She proposes changes in 'the ways in which we teach and think about our relationship with nature'. To this end, she recommends the development of a less reductionist science which allows space for ideas about the 'acquisition of gender' and which acknowledges a more holistic approach. In molecular biology, every explanation need not be reduced to the activity of DNA, but instead cellular and molecular activity can be connected. Such changes within science have to be linked to a shift in the perception of science's relation to other forms of knowledge and experience. The overall effect of such moves is intended to alter the method, aims and status of science, and thereby direct it towards a set of values and interests which are excluded at present.

If there is a common theme to all these proposals and ideas, it is that we should err on the side of caution. Our response to technology should be one of wariness, to minimise risks, to moderate effects. Adopting such an attitude is in keeping with the general approach adopted here, which emphasises the complexity of the relationship between technology and politics. Ronald Beiner argues that since we can never be certain of the consequences of our actions, we need to restrict our endeavours accordingly. He quotes Hans Jonas' injunction that 'the prophecy of doom is be given greater heed than the prophecy of bliss' (Beiner, 1988, p. 339). A similar principle emerges from Goodin's (1980) analysis of risk; he concludes that risk should be reduced to a minimum wherever possible. It also informs Elster's (1983, pp. 205–6) warning of the costs of irreversibility in decisions about technology. These writers share the view that decisions about technology have to incorporate a set of political judgements, and that those judgements should be made with due respect to the conditions of ignorance, uncertainty and complexity which pervade our relationship with technology.

CONCLUSION

In the course of this book, many more problems than solutions have been presented. This chapter has been no exception. Although I have sketched out an alternative approach to the politics of technology, I cannot pretend that it is adequate to the task. It does, however, establish a general way of thinking about both technology and politics, and the intimate connection between the two. At its grandest, it proposes that reflections on political principles cannot be conducted without reference to the technologies that give life to those principles. Statements about political values are also statements about their realisation in practice, and this is partly circumscribed by the technology. The right to speak is dependent upon the means available to make yourself heard.

Equally important to this general argument is the recognition that the technologies which give form to political principles are themselves shaped by the exercise of political choice and the struggle of competing political interests. The kinds of technology we have, and therefore the kinds of values and principles which can be articulated, are products of a political process, whether in their theory, their design, their implementation or their regulation.

If it is true that politics is dependent upon technology, and technology upon politics, then *Politics and Technology* is not so much a title as a way of life.

Bibliography

ADORNO, T. and M. HORKHEIMER (1979) *Dialectic of Enlightenment* (London: Verso).

ALBURY, D. and J. SCHWARTZ (1982) *Partial Progress: The Politics of Science and Technology* (London: Pluto Press).

ALLISON, L. (1975) *Environmental Planning* (London: Allen and Unwin).

AMANN, R. (1986) 'Technical progress and Soviet economic development: setting the scene', in R. Amann and J. Cooper (eds), *Technical Progress and Soviet Economic Development* (Oxford: Basil Blackwell) pp. 5–31.

ANGUS, I. and S. JHALLY (1989) *Cultural Politics in Contemporary America* (London: Routledge).

ARTERTON, F.C. (1987) *Teledemocracy* (London: Sage).

BACON, H. and J. VALENTINE (1981) *Power Corrupts* (London: Pluto Press).

BAHRO, R. (1986) *Building the Green Movement* (London: Heretic).

BARBER, B. (1984) *Strong Democracy: Participatory Politics for a New Age* (Berkeley: University of California Press).

BARKER, J. and H. DOWNING (1980) 'Word-processing and the Transformation of Patriarchal Relations of Control in the Office', *Capital and Class*, vol. 10, pp. 64–99.

BASALLA, G. (1988) *The Evolution of Technology* (Cambridge University Press).

BAUDRILLARD, J. (1988) 'The Masses: The Implosion of the Social in the Media', in *Selected Writings* (Cambridge: Polity) pp. 207–19.

BECKER, T. and S. SLATON, (1981) 'Hawaii Televote', *Political Science*, vol. 33, no. 1, pp. 52–65.

BEINER, R. (1988), 'Ethics and Technology: Hans Jonas' Theory of Responsibility', in R. Day, R. Beiner and J. Masciulli (eds) *Democratic Theory and Technological Society* (New York: M.E. Sharpe) pp. 336–54.

BELL, D. (1973) *The Coming of Post-Industrial Society* (New York: Basic Books).

BENJAMIN, W. (1970), 'The Work of Art in the Age of Mechanical Reproduction', in W. Benjamin, *Illuminations* (London: Cape) pp. 219–53.

BENN, T. (1982) *Arguments for Democracy* (Harmondsworth: Penguin).

BENN, T. (1988) *Office Without Power, 1968–72* (London: Hutchinson).

BENTON, S. and R. EDWARDS (1984) 'The "Greening" of the NF', *New Statesman*, 26 Oct. 1984, p. 16.

BEREANO, P., C. BOSE and E. ARNOLD (1985) 'Kitchen technology and the liberation of women from housework', in W. Faulkner and E. Arnold, *Smothered By Invention* (London: Pluto Press) pp. 162–81.

BIRKE, L. (1986) 'Toward Gender Equality in Science', in J. Harding, *Perspectives on Gender and Science* (London: Falmer Press) pp. 184–202.

BLOOM, A. (1987) *The Closing of the American Mind* (New York: Simon and Schuster).

BOLTER, J. D. (1986) *Turing's Man: Western Culture in the Computer Age* (Harmondsworth: Penguin).

BRAVERMAN, H. (1974) *Labour and Monopoly Capital* (New York: Monthly Review Press).

BREACH, I. (1978) *Windscale Fallout* (Harmondsworth: Penguin).

BRECHT, B. (1980) *Life of Galileo* (London: Methuen).

BRINTON, M. (1972) *The Bolsheviks and Workers' Control* (Montreal: Black Rose).

BROAD, W. J. (1985) *Star Warriors* (London: Faber & Faber).

BSSRS (1985) *TechnoCop: New Police Technologies* (London: Free Association Books).

BURNHAM, D. (1983) *The Rise of the Computer State* (London: Weidenfeld and Nicolson).

BURNHEIM, J. (1985) *Is Democracy Possible?* (Cambridge: Polity).

BUSH, C. (1983) 'Women and the Assessment of Technology', in J. Rothschild (ed.) *Machina Ex Dea* (Oxford: Pergamon) pp. 151–70.

CAMPBELL, D. and S. CONNOR (1986) *On the Record* (London: Michael Joseph).

CAWSON, A. (1982) *Corporatism and Welfare* (London: Heinemann).

CAWSON, A. (1989) 'European Consumer Electronics: Corporate Strategies and Public Policy', in M. Sharp and P. Holmes (eds) *Strategies for New Technologies: Case Studies from Britain and France* (London: Philip Allan) pp. 56–79.

CHANAN, M. (1988) 'Piano Studies: On Science, Technology and Manufacture from Harpsichords to Yamahas', *Science as Culture*, vol. 3, pp. 54–91.

CLARK, N. (1985) *The Political Economy of Science and Technology* (Oxford: Basil Blackwell).

CLARKE, N. and E. RIDDELL (1987) 'Who's getting a slice of the pie in the sky', *The Listener*, 15 October, pp. 10–12.

CLARKE, R. (1976) *Technological Self-Sufficiency* (London: Faber & Faber).

COCKBURN, C. (1983) *Brothers: Male Dominance and Technological Change* (London: Pluto Press).

COLLINGRIDGE, D. (1981) *The Social Control of Technology* (Milton Keynes: Open University Press).

COLLINGRIDGE, D. (1983) *Technology in the Policy Process* (London: Frances Pinter).

COWAN, R. SCHWARTZ (1985) 'The industrial revolution in the home', in D. MacKenzie and J. Wajcman (eds) *The Social Shaping of Technology* (Milton Keynes: Open University Press) pp. 181–201.

CRENSON, M. (1971) *The Un-Politics of Air Pollution* (Baltimore: Johns Hopkins University Press).

CRICK, F. (1990) *What Mad Pursuit* (Harmondsworth: Penguin).

DAHL, R. (1956) *A Preface to Democratic Theory* (Chicago: University of Chicago Press).

DAHL, R. A. (1985) *Controlling Nuclear Weapons* (New York: Syracuse University Press).

DAVIES, P. (1984) *God and the New Physics* (Harmondsworth: Penguin).

DAWKINS, R. (1986) *The Blind Watchmaker* (London: Longmans).

DEPARTMENT OF TRADE AND INDUSTRY (1991) *Evaluation of the Alvey Programme for Advanced Information Technology* (London: HMSO).

DIAMOND, N. (1981) 'The Politics of Scientific Conceptualization', in L. Levidow and B. Young (eds) *Science, Technology and the Labour Process* (London: CSE Books) pp. 32–45.

DICKSON, D. (1974) *Alternative Technology and the Politics of Technical Change* (London: Fontana).

DOBSON, A. (1989) 'Deep Ecology', *Cogito*, Spring, pp. 41–6.

DOBSON, A. (1990) *Green Political Thought* (London, Unwin Hyman).

DOUGLAS, M. and A. WILDAVSKY (1982) *Risk and Culture* (Berkeley: University of California Press).

DOWNS, A. (1957) *An Economic Theory of Democracy* (New York: Harper & Row).

DUNLEAVY, P. and B. O'LEARY (1987) *Theories of the State* (London: Macmillan).

DUNN, J. (1979) *Western Political Theory in the Face of the Future* (Cambridge University Press).

DUNN, P. D. (1978) *Appropriate Technology* (London: Macmillan).

EASLEA, B. (1983) *Fathering the Unthinkable: Masculinity, Scientists and the Nuclear Arms Race* (London: Pluto Press).

EASLEA, B. (1986) 'The Masculine Image of Science with Special Reference to Physics: How Much does Gender Really Matter?', in J. Harding, *Perspectives on Gender and Science* (London: Falmer Press) pp. 132–58.

EISENBERG, E. (1988) *The Recording Angel* (London, Picador).

ELLIOTT, D. (1978) *The Politics of Nuclear Power* (London: Pluto Press).

ELLUL, J. (1964) *The Technological Society* (New York: Vintage).

ELSHTAIN, J. B. (1982) 'Democracy and the QUBE Tube', *The Nation*, 7–14 August, pp. 108–9.

ELSTER, J. (1983) *Explaining Technical Change* (Cambridge University Press).

ENNALS, R. (1986) *Star Wars: a question of initiative* (London: John Wiley).

ENZENSBERGER, H. M. (1976) *Raids and Reconstructions: Essays in Politics, Crime and Culture* (London: Pluto Press).

EYSENCK, H. (1971) *The IQ Argument: Race, Intelligence and Education* (New York: Library Press).

FAY, B. (1975) *Social Theory and Political Practice* (London: Allen and Unwin).

FEATHERSTONE, M. (1988) 'In Pursuit of the Postmodern: An Introduction', *Theory, Culture and Society*, vol. 5, no. 2–3, pp. 195–216.

FEYERABAND, P. (1975) *Against Method* (London: New Left Books).

FISCHHOFF, B. *et al.* (1983) *Acceptable Risk* (Cambridge University Press).

FOUCAULT, M. (1980) *Power/Knowledge* C. Gordon (ed.) (Brighton: Harvester).

FOUNTAIN, N. (1988) *Underground: The London Alternative Press 1966–74* (London: Comedia).

FREEMAN, C. (1987) 'The Case for Technological Determinism', in R. Finnegan *et al.*, *Information Technology: Social Issues* (London: Hodder & Stoughton) pp. 5–18.

GALBRAITH, J. K. (1974) *The New Industrial State* (Harmondsworth: Penguin).

GARNHAM, N. (1986) 'The Media and the Public Sphere', in P. Golding, G. Murdock and P. Schlesinger, *Communicating Politics* (Leicester University Press) pp. 37–53.

GLEICK, J. (1988) *Chaos* (London: Cardinal).

GLENNY, M. (1988) 'Perestroika and the personal computer', *New Scientist*, 11 Feb., pp. 28–9.

GLOVER, J. (1984) *What Sort of People Should There Be?* (Harmondsworth: Penguin).

GOLDHABER, M. (1986) *Reinventing Technology: policies for democratic values* (London: Routledge).

GOODIN, R. (1990) 'Liberalism and the Best-Judge Principle', *Political Studies*, vol. 38, no. 2, June, pp. 181–95.

GOODWIN, A. (1990) 'Sample and Hold: Pop Music in the Digital Age of Reproduction', in S. Frith and A. Goodwin, *On Record* (New York: Pantheon) pp. 258–73.

GORZ, A. (1978) 'Technology, Technicians and Class Struggle', in A. Gorz (ed.), *The Division of Labour* (Brighton: Harvester).

GORZ, A. (1980) *Ecology as Politics* (Boston: South End Press).

GOUDSMIT, S. (1985) 'The Gestapo in Science', in M. Gardner (ed.) *The Sacred Beetle* (Oxford University Press) pp. 349–67.

GOULD, S. J. (1984) *The Mismeasure of Man* (Harmondsworth: Penguin).

GOULD, S. J. (1991) *Wonderful Life* (Harmondsworth: Penguin).

GREEN PARTY (1987) *Election Manifesto* (London: Green Party/Heretic Books).

GREENAWAY, J., S. SMITH and J. STREET (1992), *Deciding Factors in British Politics* (London: Routledge).

GREENBERG, D. (1967) *The Politics of Pure Science* (New York: New American Library).

GREY, W. (1986) 'A Critique of Deep Ecology', *Journal of Applied Philosophy*, vol. 3, no. 2, pp. 211–6.

GUSTAFSON, T. (1979) 'Environmental Disputes in the USSR', in D. Nelkin (ed.) *Controversy: Politics of Technical Decisions* (London: Sage) pp. 69–86.

HABERMAS, J. (1971) *Towards a Rational Society* (London: Heinemann).

HACKER, S. (1983) 'Mathematization of Engineering: Limits on Women and the Field', in J. Rothschild (ed.) *Machina Ex Dea* (Oxford: Pergamon) pp. 38–58.

HALES, M. (1982) *Science or Society?* (London: Pan).

HALL, P. (1981) *Great Planning Disasters* (Harmondsworth: Penguin).

HAM, C. and B. JENNETT (1987) *The Assessment and Use of Health Care Technology* (London: King's Fund Institute).

HAUG, W. F. (1986) *Critique of Commodity Aesthetics* (Cambridge: Polity).

HEILBRONER, R. (1972) 'Do Machines Make History?', in M. Kranzenberg and W. H. Davenport (eds) *Technology and Culture* (New York: Meridan) pp. 28–40.

HELD, D. (1987) *Models of Democracy* (Oxford: Polity).

HELLERSTEIN, D. (1987) 'Overdosing on Medical Technology', in A. P. Iannone (ed.) *Contemporary Moral Controversies in Technology* (Oxford: OUP) pp. 141–5.

HENDERSON, P. (1977) 'Two British Errors: Their Probable Size and Some Possible Lessons', *Oxford Economic Papers*, July, pp. 159–205

HILLMAN, M. *et al.* (1991) *One False Move: A Study of Children's Independent Mobility* (London: Policy Studies Institute).

HIRSCH, F. (1977) *Social Limits to Growth* (London: RKP).

HIRSCHMAN, A. (1982) *Shifting Involvements: private interests and public action* (Oxford: Martin Robertson).

HODGES, A. (1985) *Alan Turing: The Enigma of Intelligence* (London: Counterpoint).

HOLLICK, M. (1987) 'What is appropriate technology?', in A. P. Iannone (ed.) *Contemporary Moral Controversies in Technology* (Oxford: OUP) pp. 286–98.

HOLMES, P. and M. SHARP (1989) 'The State: Captor or Captive?', in M. Sharp and P. Holmes (eds) *Strategies for New Technologies: Case Studies from Britain and France* (London: Philip Allan) pp. 1–18.

HOSOKAWA, S. (1984) 'The Walkman effect', *Popular Music*, vol. 4, pp. 165–80.

HUGHES, T. P. (1983) *Networks of Power* (Baltimore: Johns Hopkins Press).

HUGHES, T. P. (1987) 'The Evolution of Large Technological Systems', in W. Bijker *et al.* (eds) *The Social Construction of Technological Systems* (Cambridge, Mass: MIT Press) pp. 51–82.

HUWS, U. (1988) 'Consuming Passions', *New Statesman and Society*, 19 Aug., pp. 31–4.

IANNONE, A.P. (ed.) (1987) *Contemporary Moral Controversies in Technology* (Oxford University Press).

ILLICH, I. (1975) *Tools for Conviviality* (London: Fontana).

INCE, M. (1986) *The Politics of British Science* (Brighton: Wheatsheaf).

INGLEHART, R. (1977) *The Silent Revolution* (Princeton University Press).

IRVINE, J. and B. MARTIN (1986) 'Women in Radio Astronomy – Shooting Stars?', in J. Harding, *Perspectives on Gender and Science* (London: Falmer Press) pp. 80–102.

JAMESON, F. (1985) 'Postmodernism and Consumer Society', in H. Foster (ed.) *Postmodern Culture* (London: Pluto Press) pp. 111–25.

JAMISON, A. (1989) 'Technology's Theorists: Conceptions of Innovation in Relation to Science and Technology Policy', *Technology and Culture*, vol. 30, pp. 505–33.

JASTROW, R. (1978), 'Toward an Intelligence Beyond Man's', *Time*, 20 Feb., p. 53.

JAYAWEERA, N. (1987) 'Communications Satellites: a Thirld World Perspective' in R. Finnegan *et al.* (eds) *Information Technology: Social Issues* (London, Hodder & Stoughton) pp. 195–208.

JENNETT, B. (1986) *High Technology Medicine: Benefits and Burdens* (Oxford University Press).

JENSEN, A. R. (1969) 'How much can we boost IQ and scholastic achievement?', *Harvard Educational Review*, vol. 33, pp. 1–123.

JOHNSON, E. A. (1986) 'On Being a Scientist', in J. Harding, *Perspectives on Gender and Science* (London: Falmer Press) pp. 103–9.

KALDOR, M. (1983) *The Baroque Arsenal* (London: Abacus).

KEANE, J. (1991) *The Media and Democracy* (Cambridge: Polity).

KELIHER, D. (1990) 'Core executive decision-making on high technology issues: the case of the Alvey report', *Public Administration*, vol. 68, pp. 61–82.

KEMP, R., T. O'RIORDAN and M. PURDUE (1984) 'Investigation as Legitimacy: the Maturing of the Big Public Inquiry', *Geoforum*, vol. 15, no. 3, pp. 477–88.

KENNEY, M. (1986) *Biotechnology: The University Industrial Complex* (New Haven: Yale University Press).

KUHN, T. S. (1970) *The Structure of Scientific Revolutions*, 2nd edition (University of Chicago Press).

LAING, D. (1991) 'A voice without a face: popular music and the phonograph in the 1890s', *Popular Music*, vol. 10, no. 1, pp. 1–9.

LANDES, D. (1969) *The Unbound Prometheus* (Cambridge University Press).

LANDES, D. (1983) *Revolution in Time: clocks and the making of the modern world* (Cambridge, Mass: Harvard University Press).

LAUBER, V. (1977–8) 'Ecology Politics and Liberal Democracy', *Government and Opposition*, pp 199–217.

LEE, S. (1990) *Freedom of Speech* (London: Faber & Faber).

LEWIS, P. M. and J. BOOTH (1989) *The Invisible Medium* (London: Macmillan).

LLOYD, A. (1983) 'Europe examines electronic democracy', *New Scientist*, 2 June, pp. 634–5.

LOWE, P. and J. GOYDER (1983) *Environmental Groups in Politics* (London: George Allen & Unwin).

LYON, D. (1988) *The Information Society* (Cambridge: Polity).

MACINTYRE, A. (1981) *After Virtue* (London: Duckworth).

MACKENZIE, D. (1990) *Inventing Accuracy: A Historical Sociology of Nuclear Missile Guidance* (Cambridge, Mass: MIT Press).

MAGRIS, C. (1990) *Danube* (London: Collins Harvill).

MANSBRIDGE, J. (1980) *Beyond Adversary Democracy* (New York: Basic Books).

MARCUSE, H. (1941) 'Some implications of modern technology', *Studies in Philosophy and Social Science*, vol. 9, pp. 414–39.

MARCUSE, H. (1968) *One Dimensional Man* (London: Sphere).

MARGLIN, S. (1978) 'What do Bosses do? The Origins and Functions of Hierarchy in Capitalist Production', in A. Gorz (ed.) *The Division of Labour* (Brighton: Harvester).

MARSH, M. (1985) *The Space Business* (Harmondsworth: Penguin).

MARX, K. (1954) *Capital*, Volume 1 (London: Lawrence & Wishart).

MARX, K. (1975) Preface to *A Contribution to the Critique of Political Economy*, in *Early Writings* (Harmondsworth: Penguin) pp. 424–8.

MARX, K. and F. ENGELS (1967) *The Communist Manifesto* (Harmondsworth: Penguin).

McLEAN, I. (1986) 'Mechanisms for Democracy' in D. Held and C. Pollitt, *New Forms of Democracy* (London: Sage) pp. 135–57.

McLEAN, I. (1989) *Democracy and the New Technology* (Cambridge: Polity).

McNEILL, W. H. (1983) *The Pursuit of Power* (Oxford: Basil Blackwell).

MEDAWAR, P. (1984) *Pluto's Republic* (Oxford University Press).

MEDVEDEV, Z. (1980) *Nuclear Disaster in the Urals* (New York: Vintage).

MIDDLETON, R. (1990) *Studying Popular Music* (Milton Keynes: Open University Press).

MIDGLEY, M. (1980) *Beast and Man* (London: Methuen).

MILL, J. (1937) *An Essay on Government* (Cambridge University Press).

MILL, J. S. (1972) *On Liberty and Considerations on Representative Government*, Everyman edition (London: Dent).

MILLER, D. (1983) 'The competitive model of democracy', in G. Duncan (ed.) *Democratic Theory and Practice* (Cambridge University Press).

MILLSTONE, E. (1989) 'The Regulatory Environment: Science and Politics in the Control of Technology', in M. Sharp and P. Holmes (eds) *Strategies for New Technologies: Case Studies from Britain and France* (London: Philip Allan) pp. 186–211.

MOORE, B. M. (1973) *Social Origins of Dictatorship and Democracy* (Harmondsworth: Penguin).

MUMFORD, L. (1972) 'Authoritarian and Democratic Technics', in M. Kranzenberg and W. Davenport (eds) *Technology and Culture* (New York: New American Library) pp. 50–9.

NEGRINE, R. (1989) *Politics and the Mass Media in Britain* (London: Routledge).

NELKIN, D. (1979) 'Science, Technology and Political Conflict', in D. Nelkin (ed.) *Controversy* (London: Sage) pp. 9–22.

NELKIN, D. (1987) *Selling Science: How the press covers science and technology* (New York: W.H.Freeman).

OECD (1988) *Biotechnology and the Changing Role of Government* (Paris: OECD).

OLSON, M. (1971) *The Logic of Collective Action: public goods and the theory of groups* (Cambridge, Mass: Harvard University Press).

O'RIORDAN, T. (1977) 'Environmental Ideologies', *Environment and Planning*, vol. 11, no. 1, pp. 3–14.

O'RIORDAN, T. (1983) 'The Cognitive and Political Dimensions of Risk Analysis', *Journal of Evironmental Psychology*, vol. 3, pp. 345–53.

O'RIORDAN, T. (1990) 'Global Warning', *Marxism Today*, July, pp. 12–15.

O'RIORDAN, T. (1991) 'Towards a Vernacular Science of Environmental Change', in L. Roberts and A. Weale (eds) *Innovation and Environmental Risk* (London: Belhaven) pp. 149–62.

O'RIORDAN, T., R. KEMP and M. PURDUE (1988) *Sizewell 'B': anatomy of an inquiry* (London: Macmillan).

PACEY, A. (1983) *The Culture of Technology* (Oxford: Basil Blackwell).

PARKIN, S. (1989) *Green Parties: an international guide* (London: Heretic).

PARROTT, B. (1983) *Politics and Technology in the Soviet Union* (Cambridge, Mass: MIT Press).

PATTERSON, W. (1976) *Nuclear Power* (Harmondsworth: Penguin).

PAULOS, J. P. (1990) *Innumeracy* (Harmondsworth: Penguin).

PERUTZ, M. (1991) *Is science necessary?* (Oxford University Press).

PETERSON, J. (1989) 'Eureka and the Symbolic Politics of High Technology', *Politics*, vol. 9, no. 1, April, pp. 8–13.

PINCH, T. and W. BIJKER (1987) 'The Social Construction of Facts and Artifacts', in W. Bijker, T. Hughes and T. Pinch (eds) *The Social Construction of Technological Systems* (Cambridge, Mass: MIT Press).

POOL, I. (1983) *Technologies of Freedom* (Cambridge, Mass: Harvard University Press).

POOL, I. (1990) *Technologies without Boundaries* (Cambridge, Mass: Harvard University Press).

POPPER, K. (1977) *The Logic of Scientific Discovery* (London: Hutchinson).

PORRITT, J. (1984a) *Seeing Green* (Oxford: Blackwell).

PORRITT, J. (1984b) 'Britain's Growing Greens', *Marxism Today*, March, pp. 25–9.

PRICE, B. (1984) 'Pollution: 'The invisible violence'', in D. Wilson (ed.), *The Environmental Crisis* (London: Heinemann) pp. 75–92.

PRICE, D. (1965) *The Scientific Estate* (Cambridge, Mass: Harvard University Press).

REDCLIFT, M. (1986) 'Redefining the Environmental "Crisis" in the South', in J. Weston (ed.) *Red and Green: the new politics of the environment* (London: Pluto Press) pp. 80–100.

REPPY, J. (1979) 'The Automobile Airbag', in D. Nelkin (ed.), *Controversy* (London: Sage) pp. 145–58.

RHODES, R. (1988) *The Making of the Atomic Bomb* (New York: Simon & Schuster, Touchstone edition).

RICH, R. (1987) 'Politics, policy-making, and the process of reaching closure', in H. Engelhardt and A. Caplan (eds) *Scientific Controversies* (Cambridge University Press) pp. 151–67.

RICHARDS, S. (1983) *Philosophy and Sociology of Science* (Oxford: Basil Blackwell).

ROBERTS, L. and A. WEALE (eds) (1991) *Innovation and Environmental Risk* (London: Belhaven).

ROSE, H. and S. ROSE (1970) *Science and Society* (Harmondsworth: Penguin).

ROSE, H. and S. ROSE (1982) 'Making Science Socialist', *New Socialist*, March/April, pp. 26–8.

ROSE, S., L. KAMIN and R. LEWONTIN (1984) *Not in Our Genes: biology, ideology and human nature* (Harmondsworth: Penguin).

ROSENBERG, N. (1981) 'Marx as a Student of Technology', in L. Levidow and B. Young (eds) *Science, Technology and the Labour Process: Marxist Studies*, vol. 1 (London: CSE) pp. 8–31.

ROSZAK, T. (1986) *The Cult of Information* (London: Paladin).

ROTHSCHILD, J. (1983) 'Technology, Housework, and Women's Liberation', in J. Rothschild (ed.) *Machina Ex Dea* (Oxford: Pergamon) pp. 79–93.

ROUSSEAU, J-J. (1968) *The Social Contract* (Harmondsworth: Penguin).

RUDIG, W. and P. LOWE (1986) 'The "Withered" Greening of British Politics: A Study of the Ecology Party', *Political Studies*, vol. 34, no. 2, pp. 262–84.

RUDIG, W. (1990) 'Towards a "new" political science of technology', *Strathclyde Papers on Government and Politics*, no 63.

RUNCIMAN, W.G. (1989) *A Treatise on Social Theory*, vol. II, (Cambridge University Press).

RUSSELL, B. (1985) 'The Greatness of Albert Einstein', in M. Gardner (ed.) *The Sacred Beetle* (Oxford University Press) pp. 408–12.

SAKWA, R. (1989) *Soviet Politics: an introduction* (London: Routledge).

SANDBACH, F. (1980) *Environment, Policy and Ideology* (Oxford: Basil Blackwell).

SARTORI, G. (1989) 'Video-Power', *Government and Opposition*, vol. 24, no. 1, Winter, pp. 39–53.

SCHUMPETER, J. (1976) *Capitalism, Socialism and Democracy* (London: Allen & Unwin).

SCHWARZ, M. and M. THOMPSON (1990) *Divided We Stand* (London: Harvester Wheatsheaf).

SHARP, M. (1989), 'Biotechnology in Britain and France: The Evolution of Policy', in M. Sharp and P. Holmes (eds) *Strategies for New Technologies: Case Studies from Britain and France* (London: Philip Allan) pp. 119–59.

SHARP, M. and P. HOLMES (1989) *Strategies for New Technologies: Case Studies from Britain and France* (London: Philip Allan).

SIMMS, J. (1990) *Veiled in Cloud: women and the effects of the Bhopal gas disaster* (Norwich: School of Development Studies, University of East Anglia).

SKLAIR, L. (1977) 'Science, technology and democracy', in G. Boyle, D. Elliott and R. Roy (eds) *The Politics of Technology* (London: Longman) pp. 172–85.

SMALL WORLD (1990) 'Cooking Technology', no. 9, Spring/Summer.

SMITH, A. (1986) *Wealth of Nations*, Books 1–3 (Harmondsworth: Penguin).

Social Trends 21 (1991) (London: HMSO).

STEPHENS, M. (1988) *Policing: The Critical Issues* (Brighton: Harvester Wheatsheaf).

STREET, J. (1981) 'Trade Union Attitudes to Workers Control', unpublished D.Phil., University of Oxford.

STREET, J. (1986) *Rebel Rock: the politics of popular music* (Oxford: Basil Blackwell).

SYLVAN, R. (1984) 'A Critique of Deep Ecology', Part One, *Radical Philosophy*, no. 40, pp. 2–11.

TANNSJO, T. (1985) 'Against Freedom of Expression', *Political Studies*, vol. 33, no. 4, pp. 547–59.

THEBERGE, P. (1990) 'Democracy and its Discontents: the MIDI Specification', *OneTwoThreeFour*, no. 9, Autumn, pp. 12–34.

THOMAS, K. (1984) *Man and the Natural World* (Harmondsworth: Penguin).

THOMPSON, C. (1989) 'High-technology theories and public policy', *Environment and Planning C: Government and Policy*, vol 7, pp. 121–52.

THOMPSON, E. P. (1967) 'Time, Work-discipline and Industrial Capitalism', *Past and Present*, no. 38, pp. 56–97.

THOMPSON, J. (1990) *Ideology and Modern Culture* (Cambridge: Polity).

TRILLING, L. (1972), *Sincerity and Authenticity* (Oxford University Press).

WADDINGTON, C. H. (1948) *The Scientific Attitude* (Harmondsworth: Penguin).

WAINWRIGHT, H. and D. ELLIOTT (1982) *The Lucas Plan: A new trade unionism in the making?* (London: Alison and Busby).

WAJCMAN, J. (1991) *Feminism Confronts Technology* (Cambridge: Polity).

WALKER, M. (1987) *The Waking Giant* (London: Abacus).

WALLIS, R. and S. BARAN (1990) *The Known World of Broadcast News* (London: Routledge).

WALSH, V. (1980) 'Contraception: The growth of a technology', in L. Birke *et al.* (eds) *Alice Through the Microscope* (London: Virago) pp. 182–207.

WALZER, M. (1985) *Spheres of Justice* (Oxford: Basil Blackwell).

WARD, H. (1990) 'Environment Politics and Policy', in P. Dunleavy, A. Gamble and G. Peele (eds), *Developments in Britain Politics 3* (London: Macmillan) pp. 221–45.

WEERAMANTRY, C.G. (1983) *The Slumbering Sentinels* (Harmondsworth: Penguin).

WEINER, D. (1990) 'Disaster for all', *Times Literary Supplement*, Aug. 3–9, p. 815.

WEIR, D. (1987) *The Bhopal Syndrome, Pesticides, Environment and Health* (London: Earthscan).

WEIZENBAUM, J. (1984) *Computer Power and Human Reason: from judgement to calculation* (Harmondsworth: Penguin).

WHEEN, F. (1985) *Television* (London: Hutchinson).

WIENER, M.J. (1982) *English Culture and the Decline of the Industrial Spirit, 1850–1980* (Cambridge University Press).

WILKIE, T. (1991) *British Science and Politics since 1945* (Oxford: Basil Blackwell).

WILLIAMS, R. (1971) *Politics and Technology* (London: Macmillan).

WILLIAMS, R. (1980) *The Nuclear Power Decisions* (London: Croom Helm).

WILLIAMS, R. (1989), 'The EC's Technology Policy as an Engine for Integration', *Government and Opposition*, vol 24, no 2, Spring, pp. 168–76.

WILLIAMS, R. (1990a) *Notes on the Underground* (Cambridge, Mass: MIT Press).

WILLIAMS, R. (1990b), 'Introduction to the series: Science and Politics', *Government and Opposition*, vol. 25, no. 2, Spring, pp. 212–18.

WINNER, L. (1977) *Autonomous Technology* (Cambridge, Mass: MIT Press).

WINNER, L. (1985) 'Do artefacts have politics?', in D. MacKenzie and J. Wajcman (eds) *The Social Shaping of Technology* (Milton Keynes: Open University Press) pp. 26–38.

WINNER, L. (1986) *The Whale and The Reactor* (University of Chicago Press).

WOOD, F. (1987) 'The Status of Technology Assessment', in A.P. Iannone (ed.) *Contemporary Moral Controversies in Technology* (Oxford: OUP) pp. 53–63.

WORLD COMMISSION ON ENVIRONMENT AND DEVELOPMENT (1987) *Our Common Future* (Oxford University Press).

YEARLY, S. (1988) *Science, Technology and Social Change* (London: Unwin Hyman).

YORK, P. (1985) *Modern Times* (London: Futura).

YOXEN, E. (1983) *The Gene Business* (London: Pan Books).

ZUCKERMAN, S. (1971) *Beyond the Ivory Tower* (New York: Taplinger).

Index